A History of Water Engineering and Management in Yemen

Handbook of Oriental Studies

Handbuch der Orientalistik

SECTION ONE

The Near and Middle East

Edited by

VOLUME 129

The titles published in this series are listed at *brill.com/ho1*

A History of Water Engineering and Management in Yemen

Material Remains and Textual Foundations

By

Ingrid Hehmeyer

BRILL

LEIDEN | BOSTON

Cover illustration: Water flowing from a traditional well in Yemen's capital, Ṣanʿāʾ

Library of Congress Control Number: 2018964459

Typeface for the Latin, Greek, and Cyrillic scripts: "Brill". See and download: brill.com/brill-typeface.

ISSN 0169-9423
ISBN 978-90-04-38701-0 (hardback)
ISBN 978-90-04-38771-3 (e-book)

This book is printed on acid-free paper and produced in a sustainable manner.

Contents

PART III
Water and Settlement on the Coastal Plain of Southern Yemen: The Example of Ghayl Bā Wazīr

PART IV
Water-Storage Systems in the Western Highlands of Yemen: The Cisterns of al-Jabīn

PART V
Water and Religious Magic

Acknowledgements

This book is the result of extensive field research in Yemen on the human response to water scarcity. Beginning in 1984, my graduate studies focused on ancient irrigation systems on the fringes of the desert and were carried out under the umbrella of the German Archaeological Institute (Deutsches Archäologisches Institut, DAI), Ṣanʿāʾ Branch. In 1992, I became a member of the Canadian Archaeological Mission of the Royal Ontario Museum (CAM-ROM), Toronto, and shifted my interests to medieval Islamic times. Between 2001 and 2011, I directed my own field projects, which addressed both irrigated agriculture and issues of urban water management during medieval and premodern times. The political situation in Yemen deteriorated rapidly in 2011, and fieldwork has not been possible since then.

A field project is not usually an individual endeavour. I was a member of a team when working at the German Archaeological Institute and when working with the Canadian Archaeological Mission and benefited from the expertise of my fellow team members. As a project director, I put together my own teams, which included both specialists and students of archaeology, anthropology, Arabic and Islamic studies, history, geography, and civil engineering. They all contributed to the outcome of the work. Therefore, some sections of this book are written in the first-person plural; writing in the first-person singular would simply give a false impression.

Logistical support was provided by the German Archaeological Institute (Ṣanʿāʾ Branch), under the directorship of Burkhard Vogt and Iris Gerlach, and by the American Institute for Yemeni Studies (Ṣanʿāʾ), under the directorship of Christopher Edens. I worked in Yemen under a licence from the General Organization of Antiquities and Museums (GOAM, Ṣanʿāʾ), and my colleagues from GOAM helped immensely by ensuring smooth operations in the field, besides sharing insights and local contacts. Without the support of Aḥmad al-Maṣʿabī, parts of this study would not have been possible.

Fieldwork requires substantial funding. I wish to express my gratitude for generous financial assistance to the Social Sciences and Humanities Research Council of Canada, the American Institute for Yemeni Studies, the Social Fund for Development (Ṣanʿāʾ), Shaykh Khālid Buqshān (Jedda, Saudi Arabia), the Royal Ontario Museum Foundation, and Ryerson University (Toronto). Publication of this book was supported by the Royal Ontario Museum Yemen Program Fund and by the Office of the Dean of Arts at Ryerson University. I sincerely thank Eileen Reilley for preparing the maps, drawings, and photographs.

Special thanks are due to Natalie Rose, who was a wonderful and thorough copyeditor.

Among those who have contributed greatly to the content of this book, I am particularly grateful to Andrea Moritz (Berlin) and Hanne Schönig (Halle), who have read either the entire manuscript or major parts of it. I have benefited enormously from discussions with them. Abdulrahman al-Eryani (Ta'izz and Helsinki) opened my eyes to water-management concerns that I had not seen myself. Important questions would have remained unanswered without the local informants, who generously shared their knowledge with us. Their names are cited in the individual chapters. My deepest appreciation goes to Edward J. Keall, director of CAMROM, who affirmed the validity of my research interests by assigning me responsibility in his project. It was a great privilege to work in Yemen and experience the richness of the country's heritage and the warm hospitality of its people.

The anonymous reviewer appointed by Brill gave valuable criticism and helpful comments, for which I am very grateful. I would like to offer my warmest thanks to Kathy van Vliet, Acquisitions Editor at Brill, and Ellen Girmscheid, the Production Editor, whose support was indispensable for bringing this book to fruition. Finally, I wish to thank Maribel Fierro (Madrid), M. Şükrü Hanioğlu (Princeton), Renata Holod (University of Pennsylvania), and Florian Schwarz (Vienna) for accepting this book in Brill's series *Handbook of Oriental Studies*, Section 1: *The Near and Middle East*.

A Note on Transliteration

Transliteration of Arabic generally follows that of Brill's *Encyclopaedia of the Qur'ān*:

Consonants:

ʾ	d	ḍ	k
b	dh	ṭ	l
t	r	ẓ	m
th	z	ʿ	n
j	s	gh	h
ḥ	sh	f	w
kh	ṣ	q	y

Vowels:

Long vowels:	Short vowels:	Diphthongs:
ā	a	aw
ī	i	ay
ū	u	

The Arabic article (al-) takes no alphabetic precedence in the bibliography and index listings; for example, al-Nahrawālī is alphabetized under *N*. Some exceptions to this rule occur in the usage of modern personal names, where the article is also capitalized; for example, Al-Radi and Al Tikriti are alphabetized under *A*, El-Gawhary under *E*.

Figures and Tables

Figures

Tables

Photo Credits and Permissions for Reproduction

Figures 3, 4, 7, 10, 11, 13, 14, 15, 16, 19, 20, 21, 22, 23, and 27: © DAI/Ingrid Hehmeyer

Figure 6:	Adapted from Brunner 1999, 46, with permission from the author
Figures 8 and 29:	Photo by Edward J. Keall
Figure 9:	© DAI/Burkhard Vogt
Figure 18:	© RMN-Grand Palais/Art Resource, NY
Figure 26:	© The Trustees of the British Museum
Figure 28:	Reproduced with permission from the General Organization of Antiquities and Museums (GOAM), Ṣan'ā'
Figure 35:	Reproduced from Jāzim 2008, 387, with permission from the Centre Français d'Archéologie et de Sciences Sociales (CEFAS), Ṣan'ā'
Figures 49 and 55:	Photo by Elisa Fazio

Introduction

In the past, water was the pre-eminent factor limiting life in large parts of Yemen. Everybody's livelihood depended either directly or indirectly on water, from the rulers to the farmers. The need for water spurred innovative engineering schemes in order to increase local availability; this provided the basis for the economic and cultural wealth of regions that would not have prospered otherwise. Availability of water was also imperative for the proper performance of religious rituals. Water necessitated legal regulations to ensure equitable shares or to settle disputes over a scarce resource; it shaped society because cooperation and communal responsibility were prerequisites for sustainable management. The case studies in this book provide insights into these aspects of water use and the technical and managerial struggles, failures, and successes that were involved.

1 Water, Water Engineering, and Water Management

The purpose of water engineering is to supply water for both agriculture and human settlements, while water management focuses on the strategies for making good use of the available resources so that they can meet current and future needs. Water is essential for life. It is also a complex commodity. It continuously moves in its different phases between the atmosphere, the surface, and below the surface of the Earth (fig. 1). From bodies of surface water, mostly the oceans, it evaporates into the atmosphere, where it condenses and falls as precipitation back to the Earth. Part of it collects as surface runoff in streams, from where it ultimately re-enters the oceans; the other part infiltrates the soil, where it is stored in two different ways. First, soil water dampens the soil and sustains plant growth: the water is taken up by the roots, moves through the plant, and is eventually released as water vapour from the leaves into the atmosphere, a process that is called "transpiration." Transpiration is primarily a physical necessity that enables physiological processes in the plant. Second, groundwater is stored in a porous, permeable geologic stratum (overlying an impervious layer) that is saturated with water and called an "aquifer"; we refer to the upper level of this water as the "water table." Above the table, the water is classified as "soil water"; below, as "groundwater." The groundwater can be stationary or slowly moving, so that it forms an underground stream; in places

 | DOI:10.1163/9789004387713_002

where it is not level, the water will eventually resurface as a spring.[1] Because of the interconnectedness of atmospheric water, surface water, soil water, and groundwater, human intervention can have far-reaching implications. This is of particular significance in those regions of the world where fresh water is a scarce natural resource.

People have prospered in southwest Arabia since ancient times. At first sight, the natural environment may appear uninviting, even hostile. There are no permanent rivers or lakes, and major parts of southwest Arabia have an arid or semi-arid climate, implying that the available water resources are insufficient to meet the demand for water by plants, animals, and humans. As a general rule, rain-fed agriculture requires some 250–300 mm of precipitation per year.[2] Under arid conditions, however, there is too little rainfall in an average year to sustain farming. Aridity is also characterized by considerable variability in precipitation from one year to the next, and while in some years there may be adequate rain to allow a crop to grow to maturity, in most years this will not be the case. Rain-fed farming is too unreliable to support permanent settlements. In order to be dependable, agriculture requires irrigation to rectify the imbalance between a crop's demand for water and its supply. Settlements also need water year-round, including drinking water for humans and animals, water for washing, cooking, and other domestic use, and water for religious rituals. Human sensitivity to a local environment combined with technical creativity and good water-management skills can turn a site that is, at best, marginal into a thriving settlement.

Outstanding examples of such settlements existed in southwest Arabia from ancient to medieval and premodern times. Four sites (fig. 2) are presented in this book; each is the focus of one of the four core parts (I–IV). Depending on the diversity of the material, the core parts consist of one or more case studies: Mārib is the focus of Part I, Case Study 1; Zabīd and environs are discussed in Part II, Case Studies 2–5; Ghayl Bā Wazīr is the subject of Part III, Case Study 6; and al-Jabīn is covered in Part IV, Case Study 7. To appreciate what people accomplished in each of these places, we must first address the different environments in which they settled and learned to harness the available water resources.

1 For the different types of water, see Hodge 1992, 67–68; and Brutsaert 2005, 2–3.

2 Kopp 1981, 48. The actual figure depends largely on the mean annual temperature.

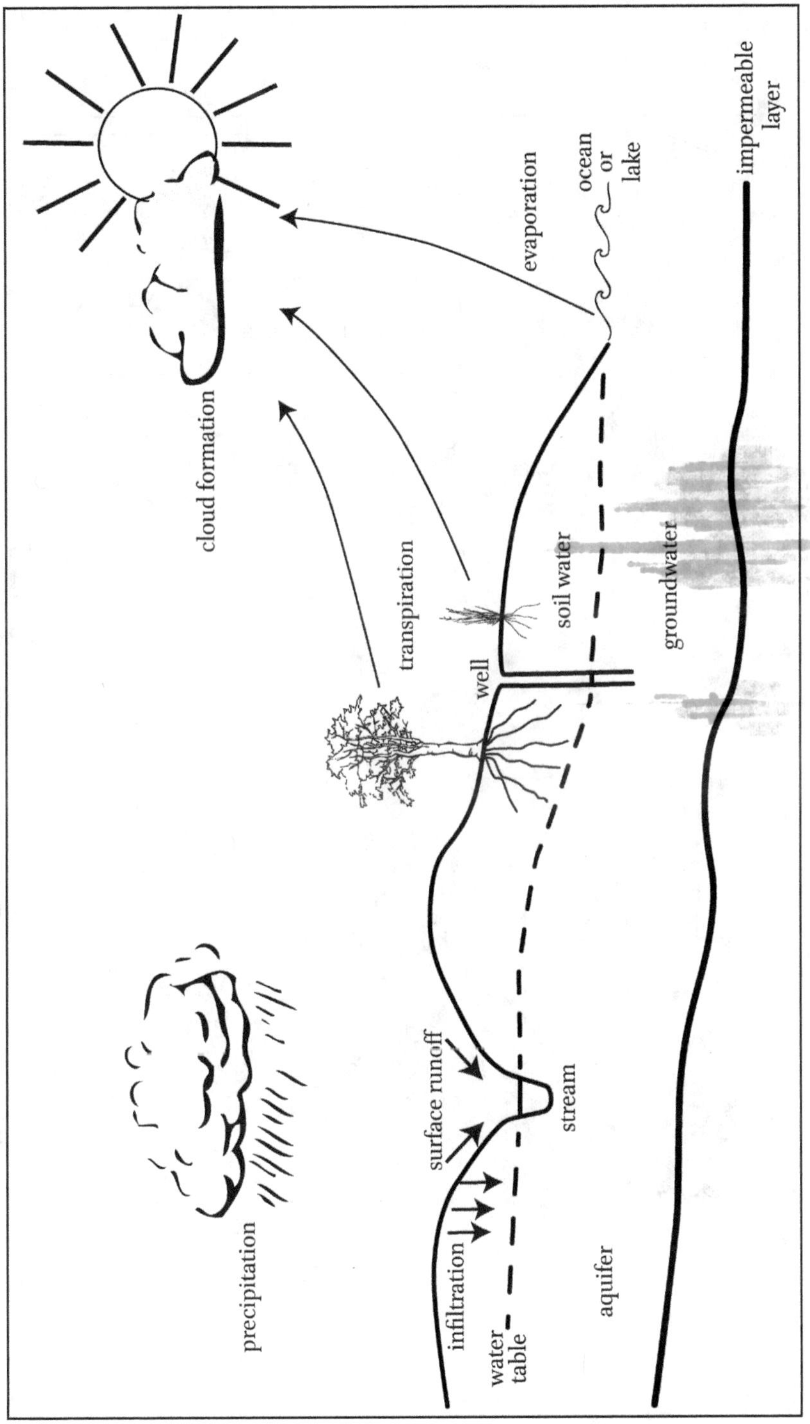

FIGURE 1 The hydrological cycle and the different types of water

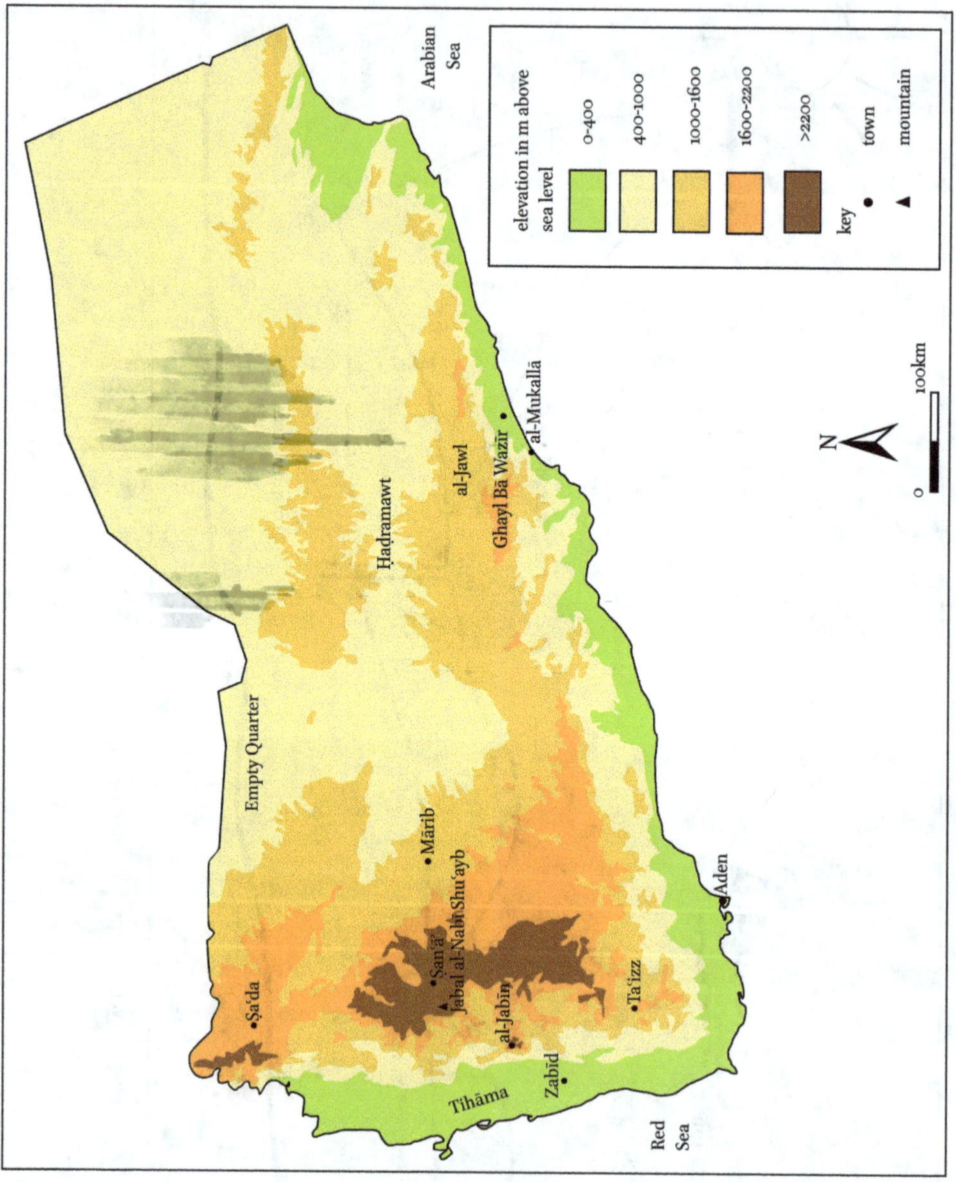

FIGURE 2 Map of Yemen showing its natural environments and places of significance to this book

2 The Natural Environments of Yemen and Their Climates[3]

The long-term averages of temperature and precipitation are what we refer to as "climate." Temperature is largely determined by closeness to the equator; the regions near the equator receive maximum energy from the sun. Yemen is located between the latitudes of 12.5 and 19 degrees north of the equator (but south of the Tropic of Cancer at 23.4 degrees north), and because of its distance from the equator, the temperature shows some seasonality. Temperature is also related to elevation above sea level; in the mountains, it is considerably lower than on the coastal plains.

There are four very different natural environments in Yemen (fig. 2).[4]

a) The country is surrounded by water on two sides: it has long coastlines both in the west, bordering the Red Sea, and in the south, bordering the Arabian Sea (which forms the northern part of the Indian Ocean). The coastal plains are mostly 20–50 km wide. The climate is hot and dry year-round, with mean annual temperatures of 30°C or more and very low precipitation, rarely exceeding 100 mm per year. Humidity is high, and dew forms regularly in the early morning when the temperature is at its lowest. Zabīd (Part II of this book) is located on the western coastal plain bordering the Red Sea, called the Tihāma. Ghayl Bā Wazīr (Part III) lies just inland from the Arabian Sea on the southern coastal plain.

b) To the east of the Red Sea coastal plain, the mountains rise steeply, with peaks of more than 3,000 m above sea level; the escarpment along the southern coast is less spectacular. At medium altitudes, the mean annual temperature is around 25°C. Precipitation is the highest in all of Yemen, averaging 500 mm or more per year; the western escarpment receives considerably heavier rainfall than the escarpment along the southern coast. On the upper escarpment and along the mountain ridge, frequent fog results in dew formation. Al-Jabīn (Part IV) is perched at the top of the western escarpment.

c) Beyond the mountain ridges, farther inland, the highlands consist of plateaus and mountain chains. The highest mountain on the Arabian Peninsula, Jabal al-Nabī Shuʿayb (3,650 m), is found in this region. Here, precipitation is below 400 mm per year, with mean annual temperatures

3 The following is based on Al-Hubaishi and Müller-Hohenstein 1984, 23–32 and 41–44; Brunner 1999, 9–24; Kopp 2008, 137–142; and Schmitz 2013, 1–12.

4 Soqotra, an island and a small archipelago in the Arabian Sea east of the Horn of Africa, is a special case and not included here. As a unique ecoregion, it would require its own investigation.

of 16°C and regular frost during the winter months. Towards the northeast, the highlands drop from some 2,500 m to 1,000 m.

d) The depression northeast of the Yemeni highlands is an extension of the desert interior of the Arabian Peninsula, the Empty Quarter (al-Rubʿ al-Khālī, the largest sand desert on Earth). Mārib (Part I) lies on the southwestern periphery of this depression. Despite the altitude, mean annual temperatures are around 25°C and precipitation is below 100 mm per year. Similar conditions prevail on the eastern plateau (approximately the eastern half of Yemen), called al-Jawl, which is essentially a stone desert. The Empty Quarter is on its north.

Precipitation is ultimately derived from the oceans. In Yemen, in simple terms, the winds draw moist air from the Indian Ocean towards the land and push it against the mountain ridges. As the air rises from a low to a high elevation under the influence of solar radiation, it cools, the moisture condenses, and it rains. Seasonal shifts in the prevailing wind systems result in two rainy seasons. Generally speaking, the rains can be expected during a shorter period in spring (March–April) and a longer period in late summer (July–August).[5] The rains typically fall as afternoon thundershowers. Precipitation is highest in the southern regions of the western highlands; the northeastern parts of the country lie in a rain shadow. This pattern illustrates the decisive role that the Yemeni mountains play in precipitation. The coastal plains receive very little rainfall, the eastern deserts even less.

It is important to note that the described seasonal pattern of rains reflects long-term averages. Localized out-of-season precipitation is not unusual, nor are dry years; every now and then, the rains will be lacking entirely. Under these conditions—low mean annual precipitation with strong fluctuations both in terms of timing and volume—average data are not truly representative and have limited significance for evaluating the agricultural potential of a specific region.[6]

The major shift towards the present climatic conditions began during the transition from the fourth to the third millennium BCE when the climate became gradually drier. This mid-Holocene climate change ended a humid period with perennial bodies of stagnant and flowing water in southwestern Arabia; lakes and rivers eventually disappeared. Apart from minor variations, Yemen's climate has remained basically unchanged for the past three millennia.[7]

5 Kopp 2005, 13–16.

6 Kopp 1981, 42–48; and Rappold et al. 2003, 71–72.

7 Brunner 1999, 25–26 and 38–39. See also Edens and Wilkinson 1998, 59–61.

3 The Water Resources

The typical watercourse of arid regions is called a "wadi." This originally Arabic term designates a dry stream bed that contains water generated by rain falling on higher ground in its catchment basin. During the rainy season, the surface runoff from the mountainsides is funnelled down and fed into the single channel of the wadi, where it results in a sudden and short-lived spate, for which the Arabic term is *sayl*. Depending on the intensity and continuance of the precipitation, a spate may last for a few hours or several days, even up to a couple of weeks in the case of recurrent thundershowers. Soon after the rains in the catchment area stop, the wadi dries up and remains dry until shortly after the rains have started again. The main characteristics of the *sayl* are its seasonality and limited duration.

At the foot of the mountains, agriculture depends on the *sayl* for irrigation; that is, the farmers use runoff from rain that falls elsewhere, higher up in the mountains. This principle is illustrated by two sites discussed in this book. Case Study 1 deals with irrigation farming in the ancient oasis of Mārib, Case Studies 2 and 3 with the medieval Wādī Zabīd. The hydrogeological conditions may afford other regional opportunities to sustain farming and settlements, such as tapping the aquifer by piped systems, as were installed in parts of the Wādī Zabīd (Case Study 4), or by gently sloping underground canals, of which Ghayl Bā Wazīr with its market gardens offers a unique example (Case Study 6). Further options to provide water for towns, including urban institutions, are open and closed canals fed by wells (and thereby groundwater), as investigated in Zabīd (Case Study 5). Use of groundwater means dependence on the height of the water table, which may fluctuate considerably over the course of a year; the potential for digging wells can be limited because of the local geology, which also affects the quality of the water under certain conditions. In some places springs exist fortuitously, but they are often seasonal. At the top of a mountain, as in al Jabīn, there is no source of fresh water apart from precipitation. Here, the surface runoff is collected and stored in cisterns (Case Study 7). Finally, under the natural conditions described above, securing the water supply is not solely a technical matter but may involve magico-religious rituals (Case Study 8).

4 A Few Words on Methodology

How do we study the history of water engineering and management? What are the sources of information, and how do we validate and interpret the evidence? A site usually catches our interest because of a written reference, a peculiarity on an aerial or satellite photograph, or because we stumble upon it by chance. Work starts with a survey of the region by way of field walking. This is essential for developing an understanding of the landscape and the natural resources, and at the same time for recording hydraulic structures (e.g., canals or water-distribution devices), remnants of construction work (e.g., plaster fragments from waterproofing a canal), or traces of irrigation farming (e.g., plough furrows). Often, archaeological excavation of a particular feature follows. Archaeology provides a more comprehensive, tangible record of the design, function, and chronology of hydraulic structures. In other cases, standing historical buildings offer the opportunity to examine construction sequences, materials, and techniques. Restoration of a historical building can further our understanding of these issues and, like archaeological excavation, may also lead to the exposure of hitherto concealed features. Written evidence from inscriptions or historical sources, where available, helps to elucidate the larger historical context and can give valuable additional information, though it is important to note that matching archaeological remains with textual statements may pose a challenge as both have unique attributes. Texts also provide details about those issues for which there are few material traces—for instance, water-management strategies. Finally, the interpretation of both the tangible and the written records can greatly benefit from the study of traditional techniques and practices that we observe today, including their terminology.[8] This involves, for example, conducting interviews with farmers and other individuals who are familiar with daily operations and challenges.[9] However, to use the present to contextualize past techniques and practices is not to perform a one-to-one transfer from the present to the past. It requires awareness of the fact that traditions are not static, and has to take into account that local informants may not necessarily give us an unbiased, complete account.

These brief remarks underscore that the history of water engineering and management is too broad and complex a topic to be addressed through a single methodological approach. It calls not only for a range of methods that allow an assessment from a variety of perspectives but also for a truly "integrated methodology"[10] in which the different approaches are interconnected and complement one another. The history of water engineering and management encourages thinking outside the disciplinary box.

---

8 For an explanation of the term "traditional," see Part II, Introduction.

9 Varisco (2015, 97–100 and 104) describes how discussions with farmers were essential for his making sense of medieval manuscripts on agriculture in Yemen.

10 This term was coined by Marcus Milwright (2010, 19) in the context of Islamic archaeology.

Each of the case studies in this book is an example of integrated methodology. The respective set of approaches used is explained in the context of the individual cases; it results from specific research questions and also from on-site circumstances, and is usually built around one core approach. For instance, because the ancient oasis of Mārib lay abandoned for some fourteen hundred years, it offered a unique opportunity to study the well-preserved traces of ancient irrigation farming during a field survey. The Wādī Zabīd, by contrast, has been farmed continuously for the past twelve hundred years and proved to be the perfect place to record techniques and practices of traditional *sayl* irrigation. A long-standing archaeological permit for Zabīd—something that should not be taken for granted—allowed us to conduct excavations in and around that city over many years. There is also a rich body of medieval sources that give invaluable details on issues such as the political and social context, investment in irrigation infrastructure, and water-allocation rights in the Wādī Zabīd. Work in Ghayl Bā Wazīr started with a field survey to record the underground canals that were still partially functioning; experienced locals shared their knowledge of the engineering process and the complex water-allocation practices with us. A mostly derelict historical cistern in al-Jabīn that we studied was subsequently restored in response to a request from the local population to ease water problems. Regional characteristics and peculiarities are particularly well reflected in the water-related technical terms that were documented in the Wādī Zabīd, Ghayl Bā Wazīr, and al-Jabīn and that are compiled in the glossary.

5 "Canal" or "Channel"?

As for water-related technical terms in English that I use throughout this book, a "canal" refers to a human-made, artificial waterway, while a "channel" designates a natural watercourse.[11] Hence, I refer to "irrigation canals" and "the flood channel of the wadi." In addition, a channel is open, while a canal may convey water in an open or closed structure. Not even the technical literature is unanimous when it comes to the terminology, and the non-technical literature often applies both terms interchangeably or uses "channel" in a human-made as well as in a natural context. Therefore, several sources quoted in this book use "channel" where "canal" would be more accurate.

11 This distinction is made, for instance, by van der Tuin 1987, nos. 584 ("canal") and 675 ("channel").

6 Brief Historical Overview

Each of the four sites has its own singular relationship to the history of Yemen that is explained in the context of the core parts of this book. In what follows, I give a brief outline of key developments and events during the prehistoric, ancient (the period of South Arabian civilization to the rise of Islam), medieval (from the rise of Islam to the first arrival of the Ottomans), and postmedieval periods. Yemen's politically and geographically complex past makes a historical categorization that applies to the whole country difficult; it has also been suggested that the medieval period extends to the eighteenth century or as late as the year 1839, when the British seized Aden.[12]

Our knowledge of the early hunter-gatherers in prehistoric southwestern Arabia is still quite limited. We gain insights into their lifestyle mostly through their stone tools and vivid rock art. The hunter-gatherers' transition into a more settled way of living, which included farming, began before the mid-Holocene climate change, when the climate was cooler and wetter. Hence, we find traces of early settlements in ecological zones where we would not expect them from today's perspective.[13]

The situation changed further when groups of people from the eastern Mediterranean started to migrate south during the second millennium BCE. At the beginning of the first millennium BCE, one of the groups of migrants, the Sabaeans, established their hegemony in South Arabia over three others: the Minaeans, the Hadramis, and the Qatabanians. The Sabaean capital and the centre of ancient South Arabia was Mārib. The earliest texts from the region that can be dated reliably are from the eighth century BCE.[14] The fifth—and last—of the ancient dynasties, the Himyarites, eventually conquered the Sabaean and neighbouring realms and unified South Arabia under their supremacy from the end of the third century CE to the sixth century. In the fourth century, monotheism increasingly replaced the ancient deities; Judaism and Christianity spread in South Arabia. One of the Himyarite rulers, who professed the Jewish faith, started to persecute the Christians. This led to the (Christian) Abyssinian conquest in 525. The Abyssinians' advocacy of Christianity in South Arabia was ended by an uprising in 570, when the local population called on the Sassanian Persian Empire for help. The Sassanians controlled South Arabia from around 575 to shortly after 628, the year of the Persian governor's conversion to Islam; Islamization of the country progressed over the course of at least

12 Caton 2013, 31 and 32.
13 Edens and Wilkinson 1998, 62–71. See also Fedele 2013, 44–50.
14 See Part I, Introduction.

three centuries.[15] In Part I of this book, I further discuss the history of ancient Mārib during the Sabaean, Himyaritic, and Abyssinian periods as it relates to the history of water engineering and management.

We have little information on early Islamic Yemen, or al-Yaman, as the southwestern part of the Arabian Peninsula is called in the Arabic sources. Prior to the establishment of Islamic dynasties in Yemen, the rulers of the Islamic Empire dispatched a number of officials to the country, including one Ibn Ziyād, who was sent by the caliph in Baghdad to settle problems in the Tihāma and founded Zabīd in 820 as his base. Ibn Ziyād eventually created the first dynasty in Yemen's Islamic history, the Ziyadids. They ruled for two centuries over territories along the Red Sea coast, on the southern coastal plain bordering the Indian Ocean, and possibly in the highlands. I deal in detail with the Ziyadid dynasty and those of its successors of significance to the medieval history of Zabīd in Part II.[16] Even though the southern coastal plain was part of the territory that several of these dynasties controlled, Ghayl Bā Wazīr's foundation and early development appear to have occurred outside Yemen's dynastic history; I explain the specific features in Part III.[17]

The tribal areas of the northern highlands, meanwhile, were ruled from the end of the ninth century onwards by (Shiite) Zaydis escaping persecution in Iraq. They created an imamate, headed by an imam, the political and spiritual leader in Shiite Islam. Over the course of a thousand years, the Zaydi imamate went through ups and downs, at times seeing strong leadership, relative political stability, economic prosperity, and the extension of its rule over large parts of the country, at other times characterized by internal squabbles, tribal anarchy, hardship, and the shrinking of their territory. The Zaydis were instrumental in liberating Yemen from the first Ottoman occupation, which lasted for almost a century (1538–1636); they subsequently became the major political force, though not uncontested. Al-Jabīn borders tribal Shiite territory, and I discuss the Zaydis' struggle for control over the troublesome tribes in the highlands in Part IV.[18] One of al-Jabīn's public cisterns originates in the post-Ottoman seventeenth century.

The second Ottoman occupation (1872–1918) was, once again, a difficult time for the Zaydis. After the withdrawal of the Ottomans, the imams enforced an isolationist policy in the northern parts of Yemen, which only came to an

15 Smith 2002, 271b–272a; and Müller 1987, 53–56.

16 See Part II, Introduction, section 2, for the historical framework; and Smith 2002, 271b–272a.

17 See Part III, Introduction and Case Study 6, section 7.

18 See Part IV, Case Study 7, section 4; and Smith 2002, 273a–274b.

end with the fall of the Zaydi imamate through the revolution of 1962. It led to the formation of the Yemen Arab Republic (YAR), or North Yemen.

In the south, the British seized Aden in 1839 in order to secure their trade routes to India; they also exercised some power over the politically fragmented hinterland, the Protectorate. Following independence from Britain in 1967, the People's Democratic Republic of Yemen (PDRY), or South Yemen, was created. The peaceful unification in 1990 of South Yemen and North Yemen resulted in the formation of the Republic of Yemen.[19] Since 2011, various tribal and sectarian groups have been vying for power, plunging the country into a civil war, and today Yemen is on the brink of becoming a failed state.

19 Burrowes 2002, 274a–275b.

PART I

Irrigated Agriculture in Ancient South Arabia: The Oasis of Mārib

Introduction to Part I

During the second millennium BCE, groups of people left the eastern Mediterranean and migrated to the southwestern periphery of the Empty Quarter. Here, they seem to have provided major impetus for the formation of an advanced civilization that developed its own material culture, religion, monumental architectural style, and writing system; the earliest reliably datable texts in Ancient South Arabian are from the eighth century BCE.[1] Ancient South Arabian is a group of related languages that were spoken (and written) in southwest Arabia between the early first millennium BCE and the rise of Islam.[2]

The people who successfully established themselves as the politically dominant force in southwest Arabia at the beginning of the first millennium BCE became known as the Sabaeans. The city of Mārib is mentioned in one of the earliest Ancient South Arabian inscriptions, but it may have emerged as the capital of the Sabaean realm, Saba', and the centre of ancient South Arabia even before writing started.[3] As one of the main stops along the caravan routes, it came to play a major role in the transhipment of incense and other aromatic substances from the areas of production around the Arabian Sea to the places of consumption in the Mediterranean region, creating considerable wealth and notoriety for the Sabaean state. The visit of the fabled Queen of Sheba to King Solomon during the tenth century BCE, which is described in both the Hebrew Bible (I Kings 10:1–13; II Chronicles 9:1–12) and the Koran (27:20–45), may have been part of such a trade mission.[4] Income from the caravan trade undoubtedly helped the state invest in infrastructure, most importantly the building of irrigation schemes that delivered even greater economic prosperity and fame: Mārib became renowned for the fertility and abundance of its irrigated farmland, the oasis of Mārib.[5] In addition to providing for the caravans, irrigation agriculture supplied the farming villages in the oasis as well as the

1 Nebes 1996, 20–22; Nebes 2001, 429–433; and Nebes 2012, 244a.

2 The terminology for the languages and scripts is explained in detail in Macdonald 2000, 30–31. It is important to note that the language, Ancient South Arabian, is capitalized. For reasons of clarity, I do not capitalize "ancient" when not referring to the language—for instance, in "ancient South Arabian civilization."

3 Müller 1991, 559b–560a.

4 Beeston 1995, 665a.

5 Gerlach (2003–2004, 38–39) points out that the incense trade played a comparatively minor role in the emergence of the ancient South Arabian states, which she attributes instead to the successful implementation of irrigated agriculture.

 DOI:10.1163/9789004387713_003

city. According to an estimate based on analogy with the traditional local farming system in the 1970s, the oasis could sustain some 30,000 people.[6] Irrigation farming in the oasis of Mārib may considerably predate the earliest epigraphic evidence, and it has been suggested that it was started by a local population whom the immigrants from the eastern Mediterranean encountered upon arrival.[7]

The city and its oasis flourished until the early centuries of the Common Era, when it became obvious that Mārib had begun to suffer from problems caused by political squabbles in South Arabia and, perhaps even more importantly, by the transfer of trade to the sea-route, which led to a decline in caravan traffic and the income derived from it.[8] Irrigated agriculture in the oasis was practised until towards the end of the sixth century CE, when the system collapsed and the site was abandoned by all but a few scattered people. The demise of the Mārib oasis is proverbial for the decline of ancient South Arabian civilization. The Koran (34:15–17) describes the devastation of the prolific gardens of Saba' and speaks of the events in moral terms, as divine punishment of people who had become too absorbed in worldly matters. However, the physical reality is that the collapse of the irrigation system reflects the fact that it had become unsustainable.[9]

New development and recultivation activities in the oasis, involving bulldozing of ancient sediments to create fields for tube-well irrigation, have resulted in extensive obliteration of the archaeological record of the field systems and ancient farming practices. However, in the late 1970s and early to mid-1980s these traces were still intact (fig. 3) and formed the basis of surface reconnaissance carried out under the umbrella of the German Archaeological Institute, Ṣanʿāʾ Branch. At the same time, the Institute began an archaeological study of the extant remains of the engineered irrigation schemes that continued until recently.[10] I was given the opportunity to participate in the November 1984–February 1985 field season in Mārib, and the results of the field research developed into a dissertation that was published in 1991.[11] Case Study 1 is an updated and extended presentation of the practices of ancient irrigated agriculture.

6 Brunner 1983, 105–106.

7 Nebes 1996, 21–22; see also Nebes 2001, 433.

8 Müller 1991, 561a–b and 564a–b.

9 The details will be discussed below in Case Study 1, section 7.

10 The results of these various field projects have been published primarily in the numerous volumes of *Archäologische Berichte aus dem Yemen* from 1982 onwards.

11 Hehmeyer and Schmidt 1991. Although German academic protocol at the time required that the program director be included as an author, the monograph is my intellectual property, with the exception of the foreword (1–8).

FIGURE 3
The ancient oasis of Mārib as seen in 1984

1 Case Methodology

Information about irrigation farming in the ancient oasis of Mārib can be derived from a variety of sources:

a) Remnants of irrigation structures allow us to understand how the water was trapped and distributed onto the fields.
b) Tangible traces of the irrigation and farming practices as well as the crops cultivated were observed in profiles exposed in the sides of gullies that were eroded down through the abandoned fields by flowing water (fig. 4). When interpreting the profiles, it is important to bear in mind that the down-cutting starts where some features are more easily eroded than others and therefore tends to follow a predictable pattern. It must

FIGURE 4 Erosion gully exposing sediment profile (1.5 m scale)

also be stressed here that the profiles were created not through archaeological excavation but through natural processes. Therefore, the relative frequency of certain features that were identified in the profiles is not necessarily representative of their original occurrence in the ancient fields.

c) Before arable cultivation was restarted on a large scale in Mārib during the second half of the 1980s, wind and occasional rain were the only forces that had affected the surface of the ancient oasis over the previous fourteen hundred years. The relief that was created from this process was not regular. As with gully erosion, minor differences in sediment constitution entailed distinctive changes, with some features being apparently more susceptible to erosion than others. This resulted in surface patterns being preserved that are visible on aerial photographs and during field surveys.[12] One can derive from them vital information about the ancient agricultural practices. However, it is important to note that, over time, progressive erosion may expose a differently patterned surface stemming from earlier times, which may lead to new conclusions.

12 It is crucial to carry out these field surveys in appropriate daylight—i.e., in the early morning or late afternoon when the sun is not too high above the horizon. In the glaring midday light, the patterns are not discernible.

d) Written evidence includes, first, monumental inscriptions in Ancient South Arabian. The texts date from the eighth century BCE to the sixth century CE.[13] As official and public communications, they were mainly carved in stone (e.g., limestone blocks or rock faces) in what is called *musnad* script. They cover a wide range of topics related to agriculture, from issues of landownership to the provision of supplies for feeding a workforce assigned to repair an irrigation structure. Despite their formal character, they give invaluable information about irrigation and farming practices in addition to details regarding the crops that were cultivated.

Second, texts in cursive handwriting, known as *zabūr*, or minuscule, script, were incised on palm-leaf stalks and wooden sticks and date to the same time as the monumental inscriptions. More informal communications, they include private letters, notes regarding transactions or delivery of goods, and private agreements, as well as contracts and other legal documents, and they enable significant insight into everyday legal and commercial life in ancient South Arabia.[14]

Due to the nature of the writing material, monumental inscriptions on limestone blocks were transported over considerable distances only in exceptional cases. Numerous inscriptions have been found in and around the Mārib oasis and provide information directly related to that area. Unfortunately, texts in minuscule script do not allow such a straightforward correlation, since a letter is typically kept in the recipient's place of residence, and identification of the place of dispatch is rarely possible. The vast majority of texts were discovered on a site some 100 km to the northwest of Mārib.[15]

The data presented here come from translations of texts and from commentaries on them. The list of such works is endless; the two most useful—and user-friendly even for the non-specialist of Ancient South Arabian—are Alexander Sima's comprehensive book *Animals, Stones, Plants, and Metals in the Ancient South Arabian Inscriptions* (2000) and Mohammed Maraqten's more recent *Texts in Ancient South Arabian on Wooden Sticks: Epigraphic and Historico-cultural Investigations* (2014). These two major works were analyzed in detail, in addition to select journal articles and book chapters. A complete and up-to-date review of all

13 Nebes 2001, 430; and Nebes 2012, 245b–246a.

14 On monumental, or *musnad*, and minuscule, or *zabūr*, script, see Maraqten 2014, 43–55; and Nebes 2012, 245b–246b.

15 See Maraqten 2014, 65–67; and Nebes 2012, 246b.

the information contained in the Ancient South Arabian inscriptions, however, would be the task of an epigrapher.[16]

e) In parts of the agricultural sector in present-day Yemen, traditional techniques and implements that have so far remained (largely) unaffected by modern influence from abroad are still being used.[17] They constitute what has been described as an "ideal laboratory"[18] for interpreting certain traces of ancient farming practices preserved in the archaeological remains.[19]

2 The Setting

Mārib's location on the southwestern periphery of the Empty Quarter means that the climate is arid. There is no substantial precipitation at all in the region,[20] and the ancient city and its hinterland could not be sustained by rain-fed farming. Instead, the imbalance between the demand for water and its supply had to be met through irrigation.

The city lies immediately to the northeast of the central highlands of Yemen, in the dry delta of the Wādī Dhana (formerly known as the Adhana wadi)[21] (fig. 5). The wadi drains an extensive catchment area in the mountains that receives significant precipitation during the biannual rainy seasons in spring and late summer. The rains result in large volumes of runoff that accumulate into spates (*suyūl*, sg. *sayl*) and flow with great force out onto the plain where Mārib is located. Generally speaking, the spate in the Wādī Dhana is fairly predictable and can be expected during a shorter period in spring (some twenty days around April) and a longer period in late summer (some thirty days around mid-July to mid-August),[22] as confirmed by a number of Ancient South Arabian inscriptions that mention the spring and late summer as irrigation seasons in Mārib.[23] Therefore, in an average year, the vegetation period between spring

16 Stein's impressive *The Minuscule Inscriptions in Ancient South Arabian on Wooden Sticks from the Bavarian State Library in Munich* (2010a) was clearly written for specialists and is not included in the following considerations.

17 For a detailed explanation of the term "traditional," see Part II, Introduction.

18 Wilkinson 2006, 38.

19 To this day, Kopp 1981 is the most comprehensive description of traditional agricultural practices in Yemen.

20 Brunner 2005, 3, fig. 3.

21 Müller 1991, 559a.

22 See Hehmeyer and Schmidt 1991, 63.

23 E.g., Gl 1679 + Gl 1773 a+b, and Gl 1762; see Müller 1983, 270–271. For another example, Ja 735 (+ Ja 754), see Müller 1986, 10–11; and Müller 1988a, 450–452. Further references are

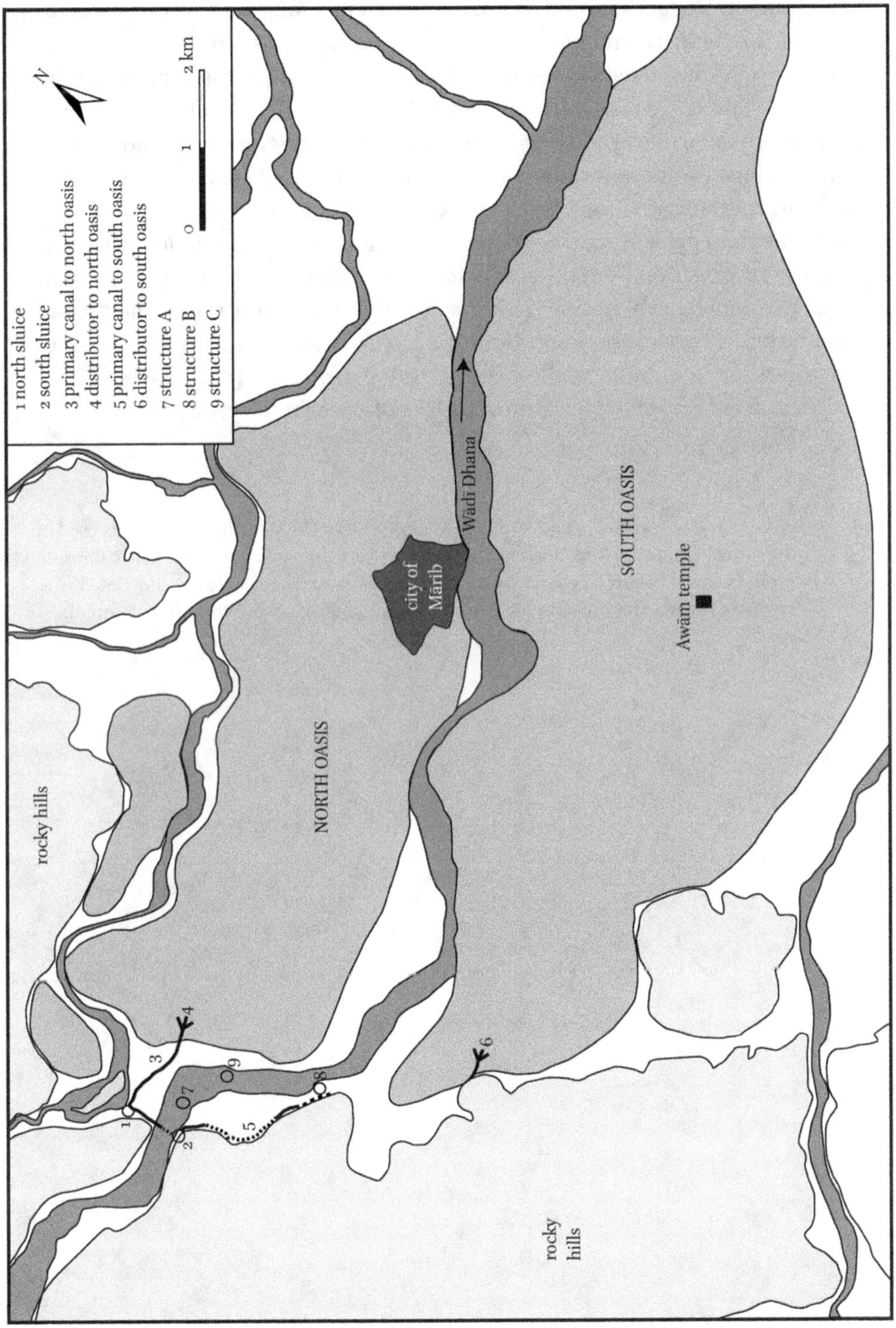

FIGURE 5 Sketch map of the Mārib oasis (adapted from Seipel 1998, 180)

and summer irrigation would have been sufficiently long to allow certain crops to grow to maturity, and parts of the oasis could have produced two harvests per year. One has to remember, though, that this pattern is based on long-term averages, and there is considerable variation in the arrival and volume of the *sayl*. More importantly, its seasonality and limited duration had fundamental implications not only for the irrigation principles employed but also for the farming methods and the choice of the crops that were cultivated. To give an example from present-day traditional *sayl* irrigation, a local sorghum variety is sown after irrigation from the spring *sayl*. During the dry months in early summer, the plants develop deep root systems that make best use of the limited soil water. A second irrigation from the late-summer *sayl* allows the crop to mature after six months. While a fast-maturing sorghum variety does exist, the farmers avoid growing it because of its lack of drought tolerance.

---

given in Hehmeyer and Schmidt 1991, 32 and nn. 66–70. The sigla that began to be used in the nineteenth century to label the Ancient South Arabian inscriptions can be confusing for those not working in epigraphy. Often, the abbreviation refers to the individual who recorded the inscription, as in the case of "Gl" (Eduard Glaser) or "Ja" (Albert Jamme).

CASE STUDY 1

Agricultural Practices in the Controlled Irrigation Network of Ancient Mārib

1 Irrigation Principles

1.1 *Diversion of the Water*

The diversion of the *sayl* required the construction of breakwaters, which were originally simple earth barrages, reaching into the wadi like fingers and directing part of the water onto the fields on either side of the flood channel (for a picture of a modern example in the Wādī Zabīd, see Case Study 2, fig. 29). Periodically, these barrages would be washed away by a *sayl* and had to be re-erected in expectation of the next spate. Because of the nature of the building material, they left no traces in the wadi. With time, engineers started to use precisely cut stone blocks as part of bigger and more robust diversion barrages to divert the water in the desired direction.[1] The engineers were also able to move the diversion devices farther and farther upstream, from where greater height for the delivery of water by gravity flow could be established.

This process eventually culminated in the construction of the famous Mārib dam, built across the mouth of the gorge where the wadi debouches from the mountains (fig. 6).[2] Despite the fact that the scholarly literature refers to it as a "dam," it was not a dam in the proper meaning of the term, designed to store water in an artificial lake or reservoir. Rather, its purpose was as an impediment to block the spate and direct the water straight onto the cultivated areas on either side of the wadi. The correct technical term for this structure—as

1 Examples are irrigation structures A and B (fig. 5), which have caused much debate with regard to their function and construction date. The issue was most recently addressed by Gerlach, who draws the convincing conclusion that they do not date to before the early first millennium BCE. See Gerlach 2012, 190–191 and 195–196. For the even more complex structure C, see Schaloske 1995, 90–92.

2 For a general description of the scheme of the Mārib dam, see Brunner 2000, 171–173. However, with regard to the architectural and functional elements of the dam as well as the attributed dating, much work has been carried out since 2002 by the German Archaeological Institute that has led to spectacular new discoveries. In this light, previously held interpretations, including the dating of the dam and its sluices, have had to be revised. See Vogt 2004b, Vogt 2005, and, in particular, Vogt 2007, for the most recent results. The irrigation network of the oasis and its function are described in detail by Schaloske (1995, 123–161).

 DOI:10.1163/9789004387713_004

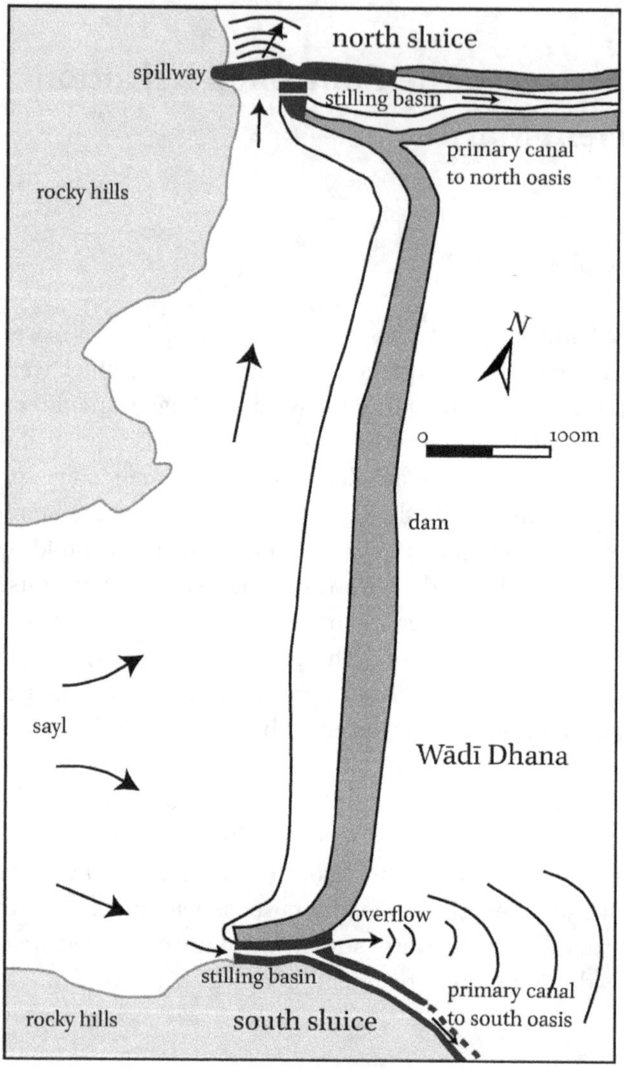

FIGURE 6
Schematic drawing of the Mārib dam, with north and south sluices

well as its more simple precursors—is a "diversion barrage."[3] In its final state, as observed and recorded by archaeologists, the central part consisted of sediments heaped up about 20 m high, with a facing of fieldstones on the upstream side to minimize water erosion (fig. 7).

The dam (as I will refer to it, following convention) spanned some 620 m across the wadi. Two massive constructions of hewn stone, known as the north and south "sluices," anchored the dam on the slopes of the mountains on

3 See Brunner 1983, 47.

FIGURE 7 Cross-section of the central part of the Mārib dam, with stone facing on the upstream side (and sediments accumulated against it)

either side, with their footings firmly set into the bedrock (fig. 8). The north sluice with two inlets controlled the flow of the *sayl* by directing it into a stilling basin that was connected to a primary canal (fig. 9). The south sluice followed the same principle but had only a single inlet through which the water was channelled into a stilling basin and from there into a primary canal. Both sluices had a spillway that allowed water above a certain gauge to flow back into the wadi, thereby controlling the volume of water entering the respective canals and protecting the irrigation infrastructure in the oasis from damage in the event of an exceptional flood.

Beyond the main division point at the end of the two primary canals, further distribution of the water under gravity flow was accomplished through a network of secondary and tertiary canals contained by earthen banks. Stone-built outlet structures allowed the water to be distributed at points of transition from a canal of higher order to canals of lower order; others served as outlets into the individual fields or as conveyance devices that regulated the discharge from one field to another. The water flow could be controlled by wooden boards that were inserted in—or removed from—corresponding grooves in the sides of the outlet openings.[4] In the early 1980s, a great number of these

4 Schaloske 1995, 133–136.

FIGURE 8 South sluice in 1987. The construction phase preserved today dates to the sixth century CE (Gerlach 2012, 189)

FIGURE 9 North sluice (with spillway, *front left*) and primary canal

structures were still preserved and scattered across the ancient oasis (see fig. 10 for an example).

Because of the limited duration of the *sayl*, the trapping, diversion, and distribution of the spate had to be completed as quickly and as efficiently as

FIGURE 10
Outlet structure

possible, so that the fullest possible extent of the cultivatable area could be irrigated. The area of farmed land was defined by the volume of water that reached the individual fields. At the time of its greatest extent, some 9,600 ha were under cultivation on either side of the wadi,[5] forming the south and north oases.

1.2 *Irrigation of the Fields*

Since water was only available during the seasonal occurrence of the *sayl*, the main aim of the operation was to deliver adequate volumes of water to the individual fields so that there would be enough moisture stored in the soil for the ensuing growing season. At Mārib, this was only possible by submerging the fields to a considerable depth, in a process best termed "ponding."

5 Brunner 1983, 90.

Accommodating a deep pond of water required the heaping up of banks around the edges of the fields. In order to appreciate how precisely the irrigation practices can be documented, one has to bear in mind that each time the fields were irrigated with particle-rich spate, a thin layer of silt was deposited. Before recultivation of the ancient oasis started to obliterate the traces from the second half of the 1980s onwards, this buildup of the sediments within a field could be observed in the aforementioned profiles exposed by gully erosion.[6]

In several of these profiles, field banks were preserved, with the sediments accumulated against them clearly distinguishable. Figure 11 shows an example from a specific moment in time during the history of the ancient oasis (see fig. 12 for a schematic drawing). The exact height of the original bank (with its top at *2*) measures 80 cm above the field surface (*1*) on which it was established and which forms the basis of the processes described here. The irrigation sediments that started to settle in the field contained by the bank can be recorded through the life of the field's use. They are composed of distinctly different strata.[7]

The bottom stratum, *a* (20 cm thick; see fig. 12), comprises the first sediments that were deposited after the field bank had been heaped up. Adjoining the bank, the sediment layers are undisturbed and each one is traceable from the respective field surface to the top of the bank. Since every layer of sediment documents one instance of irrigation, the course of the layers indicates that the field was originally submerged to the top of its bank, to the full depth of 80 cm. Over time, as sediments accumulated in the field, the submerged depth decreased to 60 cm.

Full field submersion was not the only method practised, as can be concluded from the pattern of sediment layers observed in the overlying stratum, *b* (figs. 11 and 12), where all the individual layers are pinched together at exactly the same height, some 30 cm below the top of the field bank (marked *4* in fig. 11). Clearly, each time the field was irrigated it was flooded to precisely that level. The only possible explanation for this is that a conveyance device had been built into the bank, with its threshold at a height of 30 cm below the top. It would have allowed the water to flow onto one or several fields located beyond the device. This measure may reflect a change in the sequence of water allocation.

6 The increasingly thick accumulations of sediment that retain the record of past activities have been described as "veritable agricultural museums" (Wilkinson 2002, 106).

7 For a description with full details, see Hehmeyer and Schmidt 1991, 42–46.

FIGURE 11 Field bank in profile, with sediments accumulated against it (close-up of fig. 4; 2 m scale): *1*, field surface on which the field bank was established (also the bottom of sediment stratum *a*); *2*, top of field bank; *3*, bottom of sediment stratum *b*; *4*, sediment layers of stratum *b* pinched together some 30 cm below the top of the field bank

Thus, the field was no longer submerged to the top of its bank. Instead, one may interpret the specific pattern of the sediment layers as evidence of irrigation by "controlled flooding," implying that the water did not remain fully ponded in the field but rather moved slowly in a continuous sheet of some depth while soaking the soil. Even though the depth of the water was lower than in full field submersion, the volume of water applied to the soil would have been sufficient for the ensuing growing season since the irrigation process was of longer duration.

1.3 *Supplementary (Perennial) Water Supply*

Wells and cisterns provided a supplementary perennial, though quite limited, source of water. The ancient inscriptions speak of the water stored in cisterns primarily as drinking water for animals and for domestic use.[8] Wells were dug

8 See Müller 1983, 268–269.

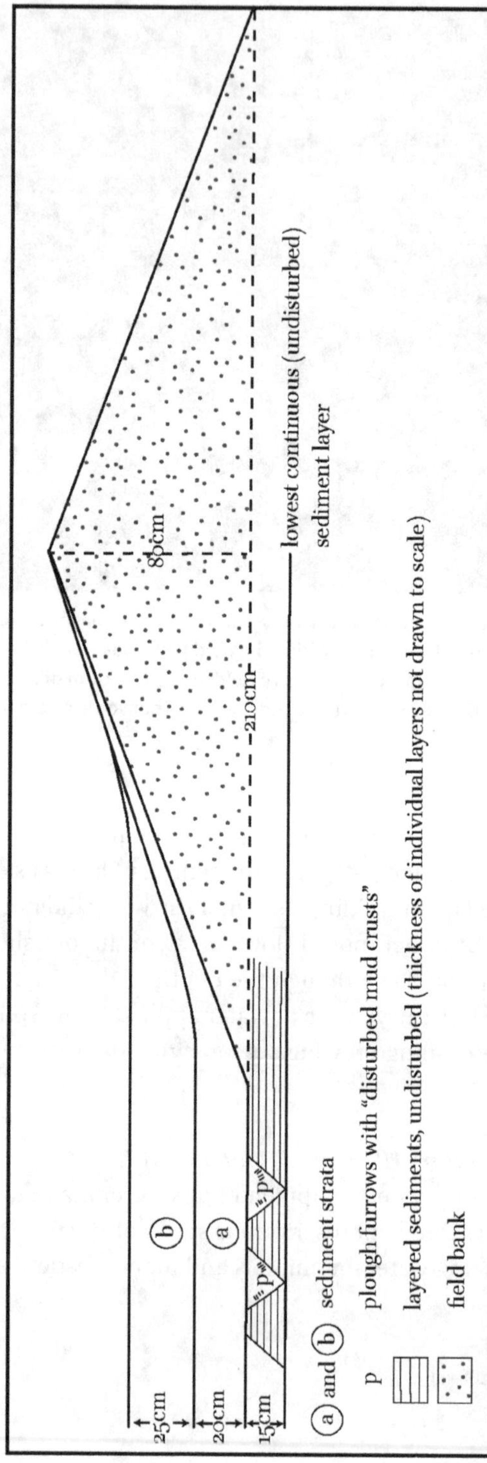

FIGURE 12 Schematic drawing of sediment profile in fig. 11

FIGURE 13
Sabaean well in the Mārib oasis, with stone lining. The arrows indicate the rope grooves, now on the underside of the rolled-over rim stone

into the aquifer to tap the groundwater and thereby provided perfectly clean water. However, dependence on groundwater meant that during times of prolonged drought, the water table dropped, and the wells dried up, a scenario described in Ancient South Arabian inscriptions.[9] A number of ancient wells have been documented in the oasis, one of which was still being used by the local Bedouin in 1984. These wells were lined with stone, a construction detail also mentioned in the written record.[10] There are clear textual indications that well water was used for irrigation.[11] Deep grooves observed in the rim stones surrounding the head of an ancient well in Mārib (fig. 13) reflect long-term use of ropes to haul buckets of water to the surface.

9 E.g., Müller 1986, 10–11; and Müller 1988a, 450–452.

10 E.g., RES 4700; see Müller 1988b, 639–640. Further examples are given in Hehmeyer and Schmidt 1991, 48.

11 E.g., Sima 2000, 224, no. 39; 224–225, no. 42; 225, no. 44; and 233, no. 81. Even though the inscriptions cited are not from Mārib, one can safely assume that the irrigation principles in the ancient oasis would have been the same.

Inscription CIS IV 570 from the Mārib area mentions a cistern, its diversion barrage, and its water distributor.[12] The text suggests that the cistern was filled by diverting water from the *sayl* and that the stored water was used for irrigation. This is confirmed by a second inscription from Mārib that reports the construction of four (covered) cisterns in a palm grove, together with their canals.[13] However, as with the use of well water, the practice of using cisterns can only imply supplementary rather than extensive irrigation of crops with high water requirements that could not be met exclusively by the periodic wadi spate, or of perennials that were dependent on a year-round water supply. It is therefore not surprising that inscriptions refer to wells and cisterns located in palm groves or vineyards.

2 Layout of the Fields

The field banks were heaped up using earth from the fields, and since the earth was composed of sediments that had been deposited through irrigation, profiles of the banks normally show small and irregular fragments of originally layered sediments that are distributed as "disturbed crusts" in the bank matrix (fig. 14). The best technical description of this formation is a "disturbed mudcrust matrix." Field banks were erected to a height of some 80–100 cm. Where a field adjoined an irrigation canal, one of the banks was shared by the canal and the field.

Evidence of earth banks is derived not only from sediment profiles in erosion gullies but also from surface patterns resulting from erosion. Undisturbed layers of sediment offer virtually no points for the erosion process to begin, which is why they are rather resistant to being worn away. A disturbed mudcrust matrix, however, as in field and canal banks, is more susceptible to the erosive forces. In the course of time, the banks could be worn down even lower than the adjoining field surfaces. Thus, linear depressions in the surface mark the original banks (fig. 15). These could be further deepened by water erosion following occasional rain.[14] The result is a rectangular erosion pattern on the

12 Müller 1983, 276–277; see Hehmeyer and Schmidt 1991, 38 and n. 79, for the translation of the term *Dammleitungskanal* as "diversion barrage."

13 Sima 2000, 227, no. 57. See also Maraqten 2014, ATHS 10, even though the origin of this text in cursive handwriting is not specified.

14 The erosion processes affecting field banks and canals are described in detail by Brunner (1983, 34–39).

surface of the oasis,[15] which is clearly discernible on aerial photographs and allows the dimensions of the ancient fields to be measured: they had an average size of 1–2 ha.

FIGURE 14 Disturbed mud-crust matrix in canal bank

FIGURE 15 Depression of eroded field bank

15 Bowen (1958, 53) coined the term "rectangular erosion" in his description of the irrigation sediments in the Wādī Bayḥān.

3 Tilling the Soil

3.1 *Ploughing*

Sediment profiles exposed through gully erosion can contain significant evidence of the techniques employed in tilling the soil. In several examples of such profiles, plough horizons were found preserved with individual plough furrows spaced 30–35 cm apart and 10–15 cm deep. Figure 16 shows a plough furrow that consists of disturbed mud crusts, while the unploughed ground is composed of layered sediments. The furrow depth of 10–15 cm corresponds to the ploughing depths obtained with the traditional wooden hook plough that is still used by Yemeni farmers today (fig. 17).[16] As the name implies, and unlike the ploughshares of modern tilling systems that turn the soil over, the hook plough simply makes grooves in the soil. This is desirable under arid conditions since it preserves moisture stored in the soil as much as possible. Today's version of the hook plough has an iron tip. From the sharp profile of the ancient plough furrows (fig. 16), one may conclude that the Sabaean plough had a similar iron tip.

The ancient South Arabian stela CIH 706, which is housed in the collection of the Musée du Louvre, Paris, and dated to the first to third centuries CE, shows a farmer ploughing with a pair of draught animals (fig. 18). In his left

FIGURE 16 Profile of plough furrow cutting through undisturbed layers of sediment (close-up of fig. 11)

16 See also Christiansen-Weniger 1971, 128–129.

FIGURE 17 Traditional Yemeni wooden hook plough, with attached sowing funnel

hand he holds the handle of the plough, while driving the animals with the whip in his right hand. The plough hook is not visible in the scene because it is running through the soil, but the attached shaft and funnel for simultaneously dropping seeds into the plough furrow are clearly depicted.[17] These technical details correspond to the plough used in Yemen today. A sowing funnel (see fig. 17) can be attached so that farmers may plough and sow in one pass, as soon as the soil is sufficiently dry after irrigation. The same sequence was practised in ancient times, which is confirmed by the sharpness of the plough furrow and its disturbed mud-crust matrix. Both features also confirm that the soil was not ploughed immediately after irrigation, when the top layer of the soil would have been wet and the plough would have "smeared" through the soil.

Evidence of ploughing is also found on the surface of the ancient oasis. Cutting into the undisturbed sediment layers by ploughing makes the now disturbed mud-crust matrix of the plough furrows more susceptible to erosion. As with the field banks, it is through selective erosion that the pattern of plough furrows is made visible on the surface.[18] One can be sure that they are ancient furrows if sediments attributable by archaeological definition to a later Sabaean period partially overlie them (fig. 19).

17 Calvet and Robin 1997, 31 and 110–111.

18 Furrows in ancient irrigation sediments of the Wādī Bayḥān were first described by Bowen (1958, 64 and plate 66) and associated with ploughing. See Brunner 1983, 31–32, for plough furrows in the ancient oasis of Mārib.

FIGURE 18 Ancient South Arabian stela (CIH 706), 1st–3rd c. CE, showing a farmer ploughing. Musée du Louvre AO 5965

A number of fields on the relict surface of the oasis in the 1980s showed a clearly visible network of plough furrows. The fields were obviously ploughed repeatedly within short intervals of time, a practice corresponding with the norms of modern irrigated arable farming, which involves ploughing the soil before irrigation in order to maximize water infiltration and to destroy weeds. As a result, the individual plough furrows were not preserved in profile. Instead, the entire plough horizon consisted of a disturbed mud-crust matrix, with only very few small fragments of layered sediments preserved.

3.2 *Fertilizing*

Sayl irrigation leads to the periodic deposition in the fields of sediments that provide nutrients to the cultivated plants. In addition, the Sabaeans employed two methods of fertilizing the soil. First, the incorporation of ashes is documented in several profiles by the presence of an ash horizon. The ash was left

FIGURE 19 Ancient plough furrows, partially overlain by sediments from later Sabaean times

behind after vegetation residues had been burnt. This measure was most probably primarily intended to remove unwanted residues and only resulted in the addition of plant nutrients as a secondary effect. When interpreting these ash horizons, one should bear in mind that the original ash layer may have become blurred through subsequent ploughing.[19]

The second method of fertilizing is documented by a dark soil horizon in the sediments, which indicates the incorporation of organic manure. Organic matter has a crucial effect on soil water, temperature, and aeration, and thus contributes in a major way to soil fertility as a whole. At Mārib, the application of organic manure would have been much needed, since even though the irrigation sediments were nutrient rich, they naturally contained little organic matter.[20]

4 Evidence of Crops

In arid regions, the question of what kinds of plants are cultivated, on the one hand, and the availability of water in terms of volume and timing, on the

19 Hehmeyer and Schmidt 1991, 27.
20 Brunner 1983, 20.

other, are interdependent. For instance, while one single deep submersion of a field from the *sayl* supplies crops such as sorghum or sesame with sufficient moisture to grow to maturity, perennials such as date palms do not survive under these conditions. The decision to cultivate them nevertheless is at the same time a decision for year-round irrigation. Thus, investigating the crops that were grown in the Mārib oasis teaches us a great deal about the irrigation practices.

The archaeobotanical record that is preserved in the sediments forms the starting point. While the traces of plants give useful hints, they allow the identification of a plant species only in exceptional cases.[21] Therefore, the tangible archaeological evidence is best complemented through information on crops furnished by the Ancient South Arabian texts. In particular, Sima's aforementioned work, *Animals, Plants, Stones and Metals in the Ancient South Arabian Inscriptions* (2000), is an invaluable source. However, as Sima points out, the plant species listed and their proportional occurrence in the texts is in no way representative of the flora of ancient South Arabia in general.[22] Monumental inscriptions were predominantly sponsored by members of the upper classes and therefore tend to reflect more precious commodities instead of everyday staples. This is why references to vineyards, for instance, by far outnumber those to barley or wheat.[23] The opposite is true of the texts in cursive handwriting on palm-leaf stalks and wooden sticks. These are of a private nature and report practical concerns of everyday life—for example, brief notes confirming delivery of a specified quantity of grain.[24] The staple crops barley, wheat, and dates predominate. It should be pointed out that—as with the monumental inscriptions—the letters, agreements, contracts, and messages were not written by the people themselves. In his *Texts in Ancient South Arabian on Wooden Sticks* (2014), Maraqten states that the documents were prepared by professional scribes according to their clients' instructions; the vast majority of the population was illiterate.[25]

---

21 Hitgen (2005a, 325, n. 6) provides preliminary identification of some archaeobotanical samples from a settlement in the Mārib oasis (8th–5th c. BCE).

22 Sima 2000, 262.

23 Interesting observations on crop preferences in prehistoric highland Yemen by Ekstrom and Edens (2003, 28–30) and Edens (2005, 193–200) should not be generalized for the Mārib oasis. The same applies to de Moulins, Phillips, and Durrani (2003), who focus on the archaeobotanical evidence from the site of Sabir (ca. 1400–850 BCE) in the Aden hinterland; the natural environment there is fundamentally different.

24 E.g., Maraqten 2014, 88–90.

25 Ibid., 27–34 and 86.

4.1 *Mud Circles*

Patterns exposed on the eroded surface of the ancient oasis give, once again, important hints. A frequently found surface feature is the mud circle.[26] It comprises three different parts (fig. 20): a circular centre with a diameter of 15–60 cm, a surrounding concentric ring up to 30 cm wide, and an outer concentric ring that is 10–50 cm wide and notable for its small and irregular fragments of layered sediments. These three zones can be interpreted based on analogy with cultivation practices observable in today's Yemen, where, in order to allow irrigation, a tree is planted in a small dug-out basin that is surrounded by a rim created by the excavated soil. In the context of ancient Mārib, this rim is represented by the outer concentric ring of the mud circle, characterized by a disturbed mud-crust matrix. The inner concentric ring corresponds to the original dug-out basin. The circular centre is a remnant of the former tree trunk, which is confirmed by the observation that, in some of the mud circles, wood residues were preserved in the centre (fig. 21). In other locations, the rim of the centre consisted of heat-affected clay. Cutting such a circle in section exposed charcoal residues resulting from the burning of the trunk, a process that left fine sediments immediately surrounding the stump as fired clay. The purpose of this practice may have been to remove the stump after felling a tree, or it may have served as a method of tree felling.[27]

Regarding irrigation, the mud circles indicate that, besides the methods described above, basin irrigation was practised in tree cultivation. Perennial crops such as trees need water year-round, and biannual *sayl* irrigation simply does not meet their water requirements. Since the trees in the ancient oasis were surrounded by individual basins that were not connected to a canal network, the water was obviously carried directly to each tree[28] so that significant losses could be avoided. The aforementioned wells and cisterns provided a

26 The first account of mud circles in Yemen was given by Bowen (1958, 60) for the Wādī Bayḥān. For Mārib, they were originally described by Gerig (1982, 48). A detailed explanation of the individual zones and their interpretation can be found in Hehmeyer and Schmidt 1991, 18–20.

27 The latter practice is described by the German cartographer and surveyor Carsten Niebuhr, who travelled in Yemen between December 1762 and August 1763 as a member of the Danish scientific expedition to the Middle East and India (1761–1767); see Niebuhr (1774) 1968, 342. Succinct summaries of the Danish expedition and its main outcomes can be found in Bidwell 1976, 32–49; and Rasmussen 1996, 57–64. For a more detailed analysis, see Baack 2014, 172–198 and 284–343.

28 Jashemski (1979, 26 with fig. 35 and 260–261 with fig. 382) describes this practice for the orchards of Pompeii, where mud circles—including the circular centre—are so well preserved in relief against the surrounding soil that the author refers to them as "sombreros."

FIGURE 20 Mud circle (1 m scale)

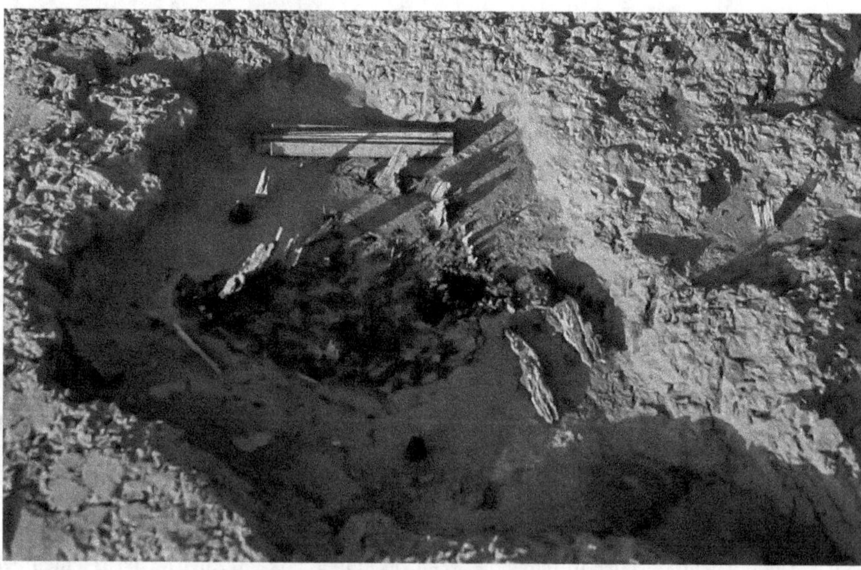

FIGURE 21 Mud circle with wood residues (20 cm scale)

perennial source of water for supplementary irrigation. However, the volume of that water was limited. This is why basin irrigation of the individual trees would have been the appropriate method to conserve the scarce supplies as much as possible.

FIGURE 22 Plough furrows deviating around a mud circle/mud mound

Two more interesting nuances can be gleaned from the surface patterns of the ancient oasis. First, mud circles were often found in straight rows, at regular intervals of 9–12 m from centre to centre, indicating deliberate rather than random planting. Second, in some instances surface traces of plough furrows show swerves around the mud circles (fig. 22), indicating that when the field was being ploughed, a tree had already been planted.[29] Thus, some of the fields where the Sabaeans cultivated trees represent examples of canopy cultivation typical of hot climates, with arable crops planted in ground shaded by mature trees.

4.2 *Fossilized Roots*

Unlike the common type of mud circle that is more or less level with the sediment surface, a second type is characterized by advanced erosion of the surrounding sediments, resulting in elevations up to 50 cm above the surface (and in a few cases even higher) (fig. 22). A better designation for this type is therefore a "mud mound."[30] In some instances, the erosion process has exposed fossilized roots in these mounds. They result from metabolic processes that occur in the outermost cell layers of a root and cement the soil particles immediately surrounding it, preserving the former root structure. Therefore, fossilized roots have the same shape as the living root. If the root is sufficiently wide, a tubular

29 Brunner 1983, 33.
30 Gerig 1982, 41 and 48.

cavity develops after the plant dies and the roots decompose. Eventually, the void may be infiltrated by sediment particles from above.[31]

4.2.1 Uniform Fossilized Roots

At Mārib, three different types of fossilized roots are found preserved in the soil. First, uniform fossilized roots occur frequently not only in mud mounds but also in sediment profiles exposed through gully erosion. They are evenly formed, approximately of the same diameter (mostly 1–2 cm) throughout their entire length, and slightly curved (fig. 23). Such root shapes are characteristic of plants belonging to the monocotyledons, a group of flowering plants whose

FIGURE 23 Uniform fossilized root

31 The development of fossilized roots is described in Glennie 1970, 113–115; and Jashemski 1979, 23.

seedlings have one seed leaf (cotyledon).[32] Since these uniform fossilized roots are often found in mud mounds or in sediment sections that have a mud circle on the surface, they clearly represent roots of trees.

The most significant family of trees within the monocotyledons are palm trees; the date palm (*Phoenix dactylifera*, Arecaceae) is, in fact, the crop most frequently mentioned in Ancient South Arabian inscriptions. The texts inform us that the Sabaeans planted extensive date-palm groves in the Mārib oasis, which was a centre of palm cultivation in ancient South Arabia.[33] As described above, according to the inscriptions, the palm groves were irrigated not only by the *sayl* but also by wells and cisterns. Year-round irrigation was, after all, a prerequisite for successful date-palm cultivation. If we follow the inscriptions, the construction and proper maintenance of the various irrigation devices were of major concern to the date farmers.[34]

The reason that the cultivation of date palms played such a major role in Mārib is their importance as a highly nutritious subsistence crop. The trees and their fruit have a multitude of uses:[35]

- Some varieties of date palms produce starchy, mealy fruits that are ground into flour and baked as bread.
- In other date varieties, the starch decomposes into sugar during the ripening process. When these fruits are dried, they are naturally preserved because of their high sugar content, making them suitable for storage. For this reason, dried dates were a staple food source for the caravans of antiquity.
- Dates can also be pressed to obtain a syrup that is either used as a sweetener or fermented into date wine.
- Date palms do not only provide food; their trunks serve as lumber for posts and roofing beams, although it is not high quality timber.
- Date-palm fronds can be used for rope making, mat weaving, and basketry.[36]

Figure 24 shows the root system of a date palm and the uniform shape of the individual roots.

---

32 The clearest description remains that in Strasburger 1978, 185, even though modern biology no longer considers monocotyledons and dicotyledons to be taxonomic categories. See Hehmeyer and Schmidt 1991, 52–53, for additional details. More recent descriptions can be found in Lüttge, Kluge, and Thiel 2010, 631 and 641–643; and Weiler and Nover 2008, 200.

33 Sima 2000, 217–239.

34 For details of date-palm cultivation in Mārib, see Hehmeyer and Schmidt 1991, 67–74.

35 Sima 2000, 239–246; for present-day uses, see Franke 1976, 268–270.

36 Date palms are pruned regularly to remove their lower fronds. Since the debris does not decompose easily, burning is the only practical method of disposal (see section 3.2 above).

FIGURE 24 Root system of the date palm (50 cm scale)

4.2.2 Non-uniform Fossilized Roots

The second type of fossilized root is bigger in cross-section than the uniform (monocotyledon) roots described above. It measures up to 15 cm in diameter and has an irregular shape. This type of root system is characteristic of the botanical group of dicotyledons, flowering plants whose seedlings have two seed leaves (cotyledons).[37] With the exception of the date palm, all the tree and shrub species that are mentioned in the Ancient South Arabian inscriptions belong to this group, the most important being Christ's thorn (*Ziziphus spina-christi*, Rhamnaceae; *ʿilb* in Arabic), a significant crop in ancient South Arabia.[38] It is the tree species mentioned second most frequently in the inscriptions. These also specify that it was cultivated in the Mārib oasis under irrigation; it did not grow randomly. Even though the texts do not go into detail regarding its various uses, one may assume that they were the same in the past as they are in today's Yemen, where *Ziziphus spina-christi* is widely cultivated for the production of an exceptionally high-quality grade of hard timber. In addition, the tree bears small cherry-sized fruits (*dawm* in Arabic) that are eaten by both humans and animals; its branches are pollarded and used as animal fodder.

37 See n. 32 above.
38 Sima 2000, 184–188.

FIGURE 25
ʿIlb tree (Christ's thorn or *Ziziphus spina-christi*)

During the particularly hot hours of the day, the tree provides shade for shepherds and their herds.[39] See figure 25 for a picture of the tree.

We also learn from the inscriptions that the ben tree (*Moringa aptera*, Moringaceae; *bān* in Arabic), is the third important cultivated tree species of ancient South Arabia, and it is specifically noted as having been planted in the oasis of Mārib.[40] In the ancient world, *Moringa aptera* was cultivated for medicinal purposes. Ben oil was obtained from its seeds and formed an important basic substance for the production of ointments that could be applied to treat skin diseases. Ben oil has been used in the Middle East until recently as lubricating grease for delicate machines like watches.[41]

39 For references to these and further uses, see Hehmeyer and Schmidt 1991, 59; and Schoenig 2006.

40 Sima 2000, 198–199.

41 Warburg 1916, 86–87; and Rhodokanakis 1931, 181–182. *Moringa* is also mentioned in a text in cursive handwriting; see Maraqten 2014, ATHS 24.

FIGURE 26 Cornice with vine scroll decoration from Mārib, 2nd c. CE. The British Museum ANE 134886

Finally, in terms of shrubs, grapevines (various species of the *Vitis* genus, Vitaceae) were cultivated in the famous Sabaean vineyards in the Mārib oasis,[42] irrigated throughout the year from wells and cisterns, as described above. It is clear that different species were grown for diverse applications, including wine making.[43] Vine scrolls are a common motif in Sabaean art, and their prevalence in architectural decoration attests to the cultural significance of the grapevine.[44] Figure 26 shows a particularly attractive example from second-century CE Mārib that is housed today in the British Museum.

4.2.3 Fine Fossilized Roots

When sediment lumps were carefully broken apart during fieldwork in Mārib, a third type of fossilized root was discovered with a very fine structure, less than 1 mm in diameter. It represents the residues of former hair roots,[45] as left by cereals and garden crops. The inscriptions mention vegetables in general ("greens")[46] and onions in particular.[47] Other garden crops such as sesame[48] and flax[49] are listed as well. The texts in cursive handwriting add various medicinal herbs.[50]

4.3 *Plant Imprints*

Imprints of crops have also been preserved in the sediments. For instance, imprints of culms with nodes and internodes (fig. 27) indicate cultivation of species from the grass family (Gramineae), economically the most important plant family since it includes the cereals. Ancient South Arabian inscriptions

42 Sima 2000, 188–196, 210–211, and 250–262.

43 On wine production and wine drinking in pre-Islamic Arabia, see Maraqten 1993.

44 For the vine scroll in ancient South Arabian art and numerous illustrations, see Costa 1992.

45 Brunner 1983, 19.

46 RES 3951, line 3; see Müller 1983, 271–273.

47 Sima 2000, 202–203.

48 Ibid., 207–209. Sesame is also mentioned in a text in cursive handwriting; see Maraqten 2014, ATHS 13.

49 Sima 2000, 216–217.

50 Maraqten 2014, ATHS 24, ATHS 25, and ATHS 31. Identification of individual species is difficult.

FIGURE 27 Sediment lump with culm imprints

list the staple crops wheat,[51] barley,[52] and (most probably) sorghum,[53] and inform us specifically that wheat and barley were cultivated in the Mārib oasis. Other imprints show the structure of pinnate leaves. These are characteristic of pulses, such as lentils.[54]

5 Evidence of Animal Husbandry and Fodder Production

The production of animal fodder was a necessity in Mārib because the Sabaeans kept large herds of animals. Draught animals were used locally not only for agricultural tasks such as ploughing but also for haulage on the engineered

51 Sima 2000, 200–202. Different wheat species are also mentioned in several texts in cursive handwriting; see Maraqten 2014, ATHS 8, ATHS 11, ATHS 17, ATHS 22, ATHS 34, and ATHS 36.

52 Sima 2000, 247–248. For barley in texts in cursive handwriting, see Maraqten 2014, ATHS 13, ATHS 15, ATHS 16, ATHS 17, ATHS 18, ATHS 19, ATHS 20, ATHS 21, and ATHS 34.

53 Sima 2000, 205–207. See Maraqten 2014, 223, for some interesting remarks on sorghum cultivation in pre-Islamic Yemen.

54 Sima 2000, 196–198. See Maraqten 2014, ATHS 35, for lentils in a text in cursive handwriting.

irrigation schemes. For instance, we learn from two well-known monumental inscriptions that the dam was destroyed in both 454 and 455 CE (CIH 540),[55] and then again less than a century later in the year 548 CE (CIH 541).[56] The texts describe the construction work on the dam and its sluices employing camels and donkeys. Detailed lists of the food provided by the ruler for the workers each time the dam and the sluices were rebuilt include slaughtered animals (cattle, sheep, and goats). The significant role that agriculture played in Mārib becomes clear in these lists of provisions: if one expects people to work hard, one has to feed them well.

Animals would also have been used to transport the provisions to the construction site. Furthermore, horses served for riding during hunting and war.[57] Above all, animal husbandry in Mārib was of utmost significance because the city was one of the stops along the incense route. The caravan trade was dependent upon healthy beasts of burden that could cope with heavy loads and long distances. Obviously, the animals needed to be well fed before departure, and well provisioned en route. As a final point, one needs to bear in mind that religious ritual in ancient South Arabia involved animal sacrifice.[58]

Pastures for animals were found in the Mārib hinterland, beyond the irrigated fields. The importance of grazing grounds to Sabaean farming is clearly expressed in inscription Ja 735 (+Ja 754), which describes a desperate situation after a prolonged drought. When the rain finally arrived, the resulting *sayl* enabled the first irrigation of the fields in a year and a half. The text then explicitly mentions that the rain also fell on the pastures.[59] Additional fresh fodder was available from pollarding trees, such as *Ziziphus spina-christi*. It is also conceivable that fodder plants were cultivated in the Mārib oasis.[60] The Sabaeans would have no doubt fed their animals weeds and all sorts of by-products and remains of crops (e.g., from sesame-oil production or threshing of cereals), and the field stubble after harvest (including the weeds) would have served for grazing. If conditions did not permit sorghum to grow to full maturity as a grain crop, it could at least be used as fodder, and immature sorghum is excellent for this purpose. All of these practices can be regularly observed in today's Yemen.

55 The complete text of the inscription together with its translation and comments can be found in Irvine 1962, 243–290. The relevant passages were checked with the help of Sima 2000, chapter 1.

56 This inscription was most recently translated by Müller (1999).

57 Sima 2000, 73–81. For further details, see Frantsouzoff 2015.

58 See Müller 1988a, for numerous examples given in inscriptions. Maraqten (2014, 313–314) comments on the significance of animal sacrifice.

59 Müller 1986, 10–11; and Müller 1988a, 450–452.

60 A text in cursive handwriting mentions the fodder plant bitter vetch (*Vicia ervilia*); see Maraqten 2014, ATHS 14.

6 Social and Legal Implications of the Irrigation Operation

6.1 *Water Allocation*

The success of the irrigation system at Mārib depended on distributing the water under gravity flow as quickly and efficiently as possible, so that the fullest possible extent of the oasis could be farmed. This was accomplished by water allocation according to the principle of upstream priority; that is, those areas farther up in the network received water before the lower-lying ones.[61] What this rule also meant was that the lower-lying field systems would not receive any water during a weak *sayl* and that parts of them could be irrigated only during the normally stronger late-summer *sayl*. The consequences are visible in the way the fields of the north oasis are stepped down sequentially, starting from the main distribution point towards the fringes.[62]

There is, however, an exception to this rule. While one would expect that the areas immediately surrounding the main division point at the end of the primary canal were always irrigated first—which would result in the sediment accumulations of those areas being highest—in the north oasis this is not entirely the case.[63] Here, the fields closest to the main division structure in fact lie lower than the area to its southeast, implying that the southeastern part of the oasis must have been given priority during water distribution. The existence of a significant settlement in this area may have been the reason.[64] Archaeological remains of houses and an enclosing wall were clearly visible in the 1980s.

An inscribed limestone stela from the region of Mārib, possibly dating to the second century BCE, records the regulation of water allocation between two adjacent estates supplied by the same canal, which meant that the estate served first had to grant the lower-lying one access to the water.[65] Inscribed on all four sides and originally painted red, with an overall height of some 80 cm[66] and a width of 24.5 cm, the stela would have been placed on the border between the two estates, to be viewed from all four sides. It was an impressive monument that not only reminded the parties concerned of the regulation but also sent

61 Upstream priority is a common practice to this day almost everywhere in the *sayl*-irrigated regions of Yemen. For details as well as for an exception to the rule, see Case Study 2.

62 Hehmeyer and Schmidt 1991, 80–82 and plate 12; and Schaloske 1995, 138–139.

63 It is not possible to make a similar judgement with respect to the south oasis because of the poorer state of preservation.

64 Schaloske 1995, 148.

65 Müller 1983, 277–278; and Mazzini and Porter 2009, 283–294.

66 The stela in its current—broken—state has a height of 43.5 cm, but reconstruction of the missing text indicates that it probably measured up to twice this height; see Mazzini and Porter 2009, 289.

a clear message to the wider community. The fact that the individual farmers would not have been able to read the text undoubtedly added authority.

This stela is not the only example of a text concerning water-allocation rights; it has a number of parallels, implying that a legal tradition regarding water was well established in ancient South Arabia.[67] We also learn from these documents that while private landownership is a characteristic feature of ancient South Arabia, the *sayl* water irrigating the land was not privately owned but shared by the community. It was therefore important to have precise rules regulating *sayl* allocation in place, and these were usually issued by a major authority or even the king himself.[68] The groundwork—that is, the management of a secondary or tertiary canal, including, first of all, water allocation, and in addition also maintenance and repair work—was the responsibility of an administrative official, the canal supervisor.[69] How stressful this task could be is spelled out in several dedicatory inscriptions. The common pattern is that "the author expresses his gratitude for having successfully accomplished his duty as the canal's supervisor. Set up at the end of his turn of office, the text reveals the author's unconcealed sigh of relief that his tenure of office passed by without any trouble."[70] Although the inscriptions never report actual fighting between farmers over water, smooth water allocation among the farmers of a canal system must have been an uphill job.

6.2 *Community Cohesion*

The situation seems to have been quite different when looking at the bigger picture, such as regular maintenance work on the primary irrigation structures. Since the periodic spate in the wadi carried silt, coarse gravel, and boulders, the sluices and main distribution structures had to be cleared from the depositions on a regular basis; otherwise, the system would have become clogged. The cleaning up following a spate is described in inscription Ja 735 (+ Ja 754), lines 14–15.[71] Another epigraphic text (dated 558 CE) reports how two tribal groups collaborated in cleanup work and removal of debris accumulated at the Mārib dam.[72]

67 For details, see Müller 1983; and Mazzini and Porter 2009, 286–287.

68 See Mazzini and Porter 2009, 289–290.

69 Stein 2010b.

70 Ibid., 340. Only one of these texts (C 365) is probably from Mārib and is dated to the second half of the second century CE, but the extant texts all reflect the same concerns.

71 Müller 1986, 10–11; and Müller 1988a, 450–452.

72 Robin 1996, 1231. The inscriptions cited are Ja 544, Ja 545, Ja 546, and Ja 547 (Christian Robin, personal communication, June 2010).

Communal responsibility and cooperation constituted the fundamental secret of how the system could remain effective for so long. In order to understand the implications, one may want to revisit inscription CIH 541 (fig. 28) and the events connected to the breaking of the dam in the year 548 CE. The text describes political problems with some of the tribes and their leaders, and how King ʾAbraha, the foreign (Abyssinian) ruler over South Arabia at the time and sponsor of the inscription, had to embark on a military operation to subdue the rebellious tribes. It was during this campaign that news reached the king that the Mārib dam had broken during the late-summer spate, in the month of July.[73] The text states that the king sent a "request" to the tribes to participate in the repairs, and an October starting date was arranged (lines 55–63). The spread of an epidemic among the workers then caused ʾAbraha to interrupt the repairs and dismiss them. But as soon as the danger had subsided, the tribesmen returned, "following his earlier request" (lines 93–94) and completed the dam in time for the spring spate.

It seems amazing that ʾAbraha's "request"[74] was followed without resistance by the insurgent tribesmen, possibly more so since ʾAbraha was a foreigner. The workers did not take the interruption to the repairs resulting from the epidemic as an opportunity to simply disappear either. If the army had had to forcibly conscript people, the text would mention it. Instead, willing participation and shared responsibility are signalled by the workers' behaviour, even though one cannot imagine that everybody was necessarily enthusiastic about working for months on a construction site. The only plausible explanation is a common understanding of the utmost significance of water in general—and of the smooth operation of Mārib's irrigation system in particular—for everybody's livelihood, whether the individual was from Mārib itself or from other parts of South Arabia. The demise of the economic viability of the Mārib oasis would have had a strong impact on the entire state and the lives of the tribespeople, and they rallied to the cause, recognizing that fact. Clearly, this attitude resulted from experiences of disasters caused by prolonged drought. Everyone suffered in these situations, and the inscriptions describe the despair and the religiously inspired

73 CIH 541, lines 41–46; see Müller 1999, 268–270, for the translation of CIH 541. For the slightly revised date of the breaking of the dam, see Nebes 2004, 229, n. 12.

74 The German term *Aufforderung* (literally, "request") is used by Müller in his translation of CIH 541. A similar term is employed by Kropp in his 1994 translation of excerpts of CIH 540, an inscription that reports the events connected with the breaking of the dam about a hundred years earlier. The relevant passage reads that the king issued a "call" (*Aufruf* in German) to participate in the repairs; see Kropp 1994, 129. In his article, Kropp gives an interesting interpretation of the passage, employing the rationale and jargon that are familiar from today's disaster management.

FIGURE 28
ʾAbraha's stela with the account on his deeds (CIH 541) from Mārib, 549 CE. Mārib Museum BAR 2

rituals that emanated from the entire community in response to a particular calamity. This issue will be addressed in detail below; see Case Study 8.

7 Technical Limitations and the Demise of the Mārib Dam

The long-term success of irrigation agriculture in Mārib was only possible because the fertility of the soil was guaranteed for two reasons. Firstly, submerging the fields with large volumes of water resulted in salts being leached into the subsoil. Problems with salinization of the top soil layers were thus avoided. Secondly, the use of particle-rich water for irrigation ensured the periodic deposition of nutrient-rich sediments in the fields.[75]

However, an enormous disadvantage of the constant deposition of sediments was the continually rising level of the field surfaces. Much has been written about the average annual sediment buildup on the oasis; the most frequently cited figure for Mārib is approximately 1 cm per year.[76] This principle is applied as a tool by which sediment accumulation can be dated "by the yardstick."[77] One must bear in mind, though, that while it may be a useful rule of thumb, the average annual sedimentation rate does not constitute a natural law. Anomalies may have occurred during the buildup of a specific sediment profile, and one should be cautious about measuring the sediments and deducing an exact date for them. In addition, the natural regime of the Wādī Dhana—before the onset of human intervention and management of the *sayl*—involved accumulation of sediments on the flood plain in the classic manner of delta formation. It is hardly possible to determine the baseline of human activities. Consequently, it is not generally appropriate to measure, for example, 10 m of sediment and conclude that this represents a thousand years of irrigation.[78]

Nevertheless, notwithstanding these cautions, the reality in Mārib was that constant irrigation resulted in the need for the irrigation structures to be raised again and again. The aim was to force the water to a height that would still allow it to be diverted onto the elevated fields under gravity flow. Eventually, the entire irrigation system from the dam and the sluices to the canals and the fields was elevated to, and operating at, a considerable artificial height. There is an inherent danger in a large volume of water being retained at

75 Brunner 2000, 175–176.

76 Brunner 2005, 4. Balescu et al. (1998, 36) calculated the annual sedimentation rate at the site of al-Ḥaraja in the Wādī Bayḥān at also about 1 cm per year, based on dating of the irrigation sediments by optically stimulated luminescence (OSL).

77 Vogt 2004a, 70.

78 See also Francaviglia 2002, 117–118, 138, and 141.

an unnatural height: that it will follow the law of gravity and find a weak spot in the system, break out of its man-made confinements and cut a path to a lower level. It is not surprising, therefore, that breaches within the irrigation system occurred. Erosion gullies cutting through the deep sediment accumulations of the ancient field systems are testimony to the phenomenon at least during the last centuries preceding the collapse of the oasis.[79]

A concomitant of the sedimentation process was the loss of stability of the dam and the sluices as a result of the constant rise in height. The north sluice can serve as an example. The structure that is preserved to this day was only completed in 548 CE.[80] Iron rods embedded in dovetail joints filled with lead were introduced to interlock the ashlars in order to provide additional strength in critical places.[81] This technique is also described in the mid-fifth-century CE inscription CIH 540 as having been employed in repairing the dam and its sluices after breaches had occurred in two consecutive years, 454 and 455 CE (lines 12–13 and 76).[82] In addition, the inscription mentions the use of lime plaster, today known as *qaḍāḍ*, on the masonry (lines 78–79).[83]

Unfortunately, disastrous breaches like the ones described in CIH 540 were becoming quite common, indicating that even the use of iron tie-rods and *qaḍāḍ* for increased stability provided only a temporary solution to a severe technical problem. From the beginning of the fourth century CE until the end of the sixth century when the final breach occurred, inscriptions refer to five incidents of the dam breaking:

- at the beginning of the fourth century CE
- at the beginning of the second half of the fourth century CE
- in 454 CE
- in 455 CE
- in 548 CE[84]

Based on differences in the cross-bedding of the sediments (what are called "discordances") observed in the entrapment area behind the dam, Ueli Brunner concludes that a sixth breach occurred towards the end of the sixth century.[85]

79 Schaloske 1995, 146–147, 157–158, and plates 18a and b. Brunner (1983, 67) had already speculated about the possibility.

80 Vogt 2005, 512.

81 Vogt 2004b, 383; and Vogt 2007, 124–126.

82 Irvine 1962, 247–251.

83 For details of the use of *qaḍāḍ* on the north sluice, see Vogt 2004b, 383; and Vogt 2005, 510–513. The composition, preparation, and application of *qaḍāḍ* are explained in Part IV, Case Study 7, sections 2 and 3.

84 Müller 1991, 563–564. See Nebes 2004, 229, nn. 10 and 12, for the slightly revised dates.

85 Brunner 1983, 51–53, 118–119, and 123.

It is obvious from interpreting the texts and from the physical reality that each time the barrage was breached, it was restored. The repairs were made with the utmost urgency in order to resume operations before the next spate season. Otherwise, no water would have been available for irrigation, and there would have been no harvest. Without its functioning irrigation system, the oasis was dead.

The final catastrophe after a disastrous flood in the wadi in the last years of the sixth century CE is described in the Koran (34:15–17a):[86]

> For Sheba also there was a sign in their dwelling-place—two gardens, one on the right and one on the left: "Eat of your Lord's provision, and give thanks to Him; a good land, and a Lord All-forgiving." But they turned away; so We loosed on them the Flood of Arim, and We gave them, in exchange for their two gardens, two gardens bearing bitter produce and tamarisk-bushes, and here and there a few lote-trees. Thus We recompensed them for their unbelief.

This destruction of the dam led to the enforced emigration of the resident population and the abandonment of the oasis by all but a few scattered people. Entire South Arabian tribes migrated north.[87] Remarkably, many of the irrigation works that they left behind were well preserved in the oasis in the mid-1980s. The archaeological study of the remains provides an awe-inspiring account of technological advances made over time, in hydraulic engineering as well as in cultivation techniques and metal manufacture. They were the result of skills developed over many centuries by people who learned to adapt to the local conditions and to make the best use of natural resources.

Whether or not these technologies were imported is currently under debate. Burkhard Vogt argues against a transfer of technologies by the immigrants from the eastern Mediterranean who started to arrive in South Arabia towards the end of the second millennium BCE. Instead, he credits the indigenous, pre-immigration population of the Mārib area as having developed the sophisticated technologies described above with no outside influence. The immigrants simply took over from them, without introducing any major innovations in technological know-how.[88] This interpretation has been challenged by Holger Hitgen, who emphasizes the significance of the technological contribution made by the newcomers. In his opinion, it was, in particular, the sophisticated stonemasonry technique that the immigrants brought with them that, com-

86 The quotation from the Koran follows Arberry 1998.

87 Müller 1991, 564a.

88 Vogt 2004a, 98.

bined with the indigenous population's age-old understanding of the scarce water resources, enabled the irrigation schemes of Mārib to be built.[89]

Vogt bases his argument on the fact that the irrigation sediments, resulting from man-made irrigation schemes, reach a maximum height of about 30 m,[90] a value that had originally been suggested by Brunner based on his measurements of sediment profiles.[91] Dating these sediments by the yardstick—that is, applying an annual sedimentation rate of 1 cm—would push the beginnings of irrigation in Mārib back to the middle of the third millennium BCE. However, Brunner himself is much more careful towards the end of his book, where he refers back to the issue as being based on the "assumption" (*Annahme* in German) that the sediments he measured were actually deposited by irrigation.[92] Even if this assumption were proven correct, the sediments should not necessarily be interpreted as resulting from monumental hydraulic engineering works that required sophisticated technological skills in stonemasonry and metalworking. One should not discount the possibility that the lower sediment accumulations in the profiles measured by Brunner might more likely be the result of natural flood-plain formation or of simple earth-diversion barrages in the wadi, as employed by the indigenous population in Mārib starting at an early date, long before the arrival of the immigrants. The conclusion drawn by Vittoria Buffa and Vogt that the Mārib dam does not reflect imported technology but the "refinement of skills developed since the Early Bronze Age"[93] also does not take into consideration that there is a fundamental technological difference between these simple barrages and the sophisticated monumental irrigation structures built from the first millennium BCE onwards. Incidentally, the practice of irrigation by simple earth-diversion barrages was retained in Sabaean times.[94] May one suggest that the rate of the annual sedimentation buildup under these conditions might differ from 1 cm per year?

Whatever the origin of the technologies employed in the irrigation structures, Mārib clearly represents a tragic example of how technology does not always have the ultimate answer. The increasingly sophisticated engineering works to manipulate the *sayl* created a problem that would ultimately cause the demise of the system. The catastrophe became inevitable because "the system had reached its technical limits."[95] Mother Nature had her way in the end.

---

89 Hitgen 2005b, 25 and 38–39.
90 Vogt 2004a, 98.
91 Brunner 1983, 65 and 74.
92 Ibid., 107–108.
93 Buffa and Vogt 2001, 446.
94 See Hehmeyer and Schmidt 1991, 37–38 and nn. 78–79.
95 Vogt 2004b, 387.

PART II

The City of Zabīd (Founded 820 CE) and its Agricultural Hinterland before the First Ottoman Conquest in 1539

Introduction to Part II

Sayl irrigation was widely practised in ancient Yemen. The pre-Islamic settlements, of which Mārib is a prominent example, are mainly found along the foot of the highlands, where major wadis debouch from the mountains onto the western and southern fringes of the desert interior of the Arabian Peninsula.[1] While the sites functioned as stops along the caravan routes, it was the control of the periodic spate in the wadis that provided the basic resources on which life depended. Application of the ancient principles of *sayl* irrigation continued after the rise of Islam at the beginning of the seventh century CE. For major parts of Yemen, the seasonal spate is simply the only source of water for anything but small-scale irrigation. For centuries, the memory of ancient Mārib and the abundance of its oasis loomed large in people's minds. A fourteenth-century reminiscence describes how "a servant would walk among the trees with a large basket on his head. The basket would fill with fruits without his picking any with his hand, and without gathering any from the ground. The sun could never reach anyone who walked beneath the trees of these gardens, by reason of their interlacing."[2]

The following four Case Studies (2–5) focus on the city of Zabīd, a small town that is located some 25 km inland from the Red Sea on the coastal plain called the Tihāma, and its agricultural hinterland in the Wādī Zabīd. This takes us into the Islamic period, since Zabīd was founded in the year 820 CE.

Unlike the ancient oasis of Mārib, which lay abandoned from around 600 CE until the mid-1980s, thus preserving the archaeological record of the ancient agricultural practices in a unique way, the Wādī Zabīd has been farmed continuously to the present. Intervention in the engineering of the irrigation structures only occurred during the late 1970s as part of the Tihāma Development Project (completed in 1979). The underlying problem making the project necessary was once again sedimentation, which had caused field levels to rise to the point where irrigation from the wadi was no longer possible. The main irrigation structures were raised and made permanent by using concrete to replace the earth-and-stone diversion barrages and primary canals. Even though some details as well as certain canal names and irrigation areas changed,[3] the

1 See Brunner and Haefner 1990, 135–152.

2 Al-Khazrajī/Redhouse, trans., and ʿAsal, ed. 1906–1918, 1:52 (Eng. trans.), 4:8 (Ar. text). See section 1 below for information on the author.

3 For instance, several barrages were constructed across the wadi, and former primary canals became subordinate ones. See Bonnenfant 1997, 17, for a map of the Wādī Zabīd after

 | DOI:10.1163/9789004387713_005

subsidiary irrigation works remained largely the same, as did the general layout of the system. As a result, the Wādī Zabīd offers a unique opportunity for first-hand observation of traditional water distribution and farming practices, and for documentation of their terminology.

Working on the Islamic period means that we have a rich body of written sources at our disposal. The challenge, then, is to interpret the archaeological remains in light of the written evidence while using the traditional traces for comparative assessment. The latter can be of great help for piecing together archaeological remains or trying to make sense of scattered information on irrigation in the written sources. Of necessity, the methodological approach in the Wādī Zabīd is different than at Mārib.

What does the term "traditional" imply?[4] Traditional knowledge and practices are accumulated over many generations by people while securing their livelihoods. They are transmitted mostly orally from the older generation to the younger, as well as through imitation and demonstration.[5] Traditional knowledge and practice are based on, and refined by, long experience. To consider them things of the past, inherited and strictly applied by a group of people, would be wrong. Instead of being static, they are adaptive, the "product of an interplay of forces,"[6] albeit so far unaffected by modern influence from abroad. It is through their historical dimension that traditional knowledge and practices link the present to the past.

Traditional water engineering and management are shaped by the local ecosystem in which they develop. Clearly, the fact that a particular technical approach or management practice has prevailed over an extended span of time attests to its sustainability. This means that it allowed one generation to meet its needs without compromising the ability of future generations to provide for themselves. Because of the complexity of ecosystems in general, and arid ecosystems in particular, the traditional technologies managing them evolved as multifaceted, integrated, and therefore flexible response systems.[7] Negative environmental effects that would have jeopardized long-term success—and

---

implementation of the Tihāma Development Project. Some further changes have taken place since then.

4 Some consider use of the term "traditional" problematic because it may be misunderstood as denoting "19th-century attitudes of simple, savage and static" (Warren 1996) and prefer "indigenous" instead. It is quite evident in the context of this book, though, that such a concern is unfounded. "Simple," "savage," and "static" misrepresent the proper sense of the term.

5 See Berkes, Colding, and Folke 2000, 1251–1252; and Ellen 2003, 65.

6 Gerholm 1984, 88.

7 See Laureano 2003, 142, table 1.

thereby the livelihood of the local community—were gradually corrected, based on people's skills and experience.

It is important to acknowledge that it is the members of a local community who have devised particular engineering or management approaches, and it is they who account for their proper execution. Traditional systems of water engineering and management therefore tell us as much about the technicalities as about the interaction of people with their environment. It is equally significant to bear in mind, however, that any gross change of the underlying parameters can have disastrous consequences when the accumulated knowhow is no longer valid. A fundamental challenge also arises when people's appreciation for their shared skills diminishes because modern technologies that are introduced from abroad appear to render them useless.

1 The Historical Framework

Historical sources such as official histories and chronologies give information on, and aid in the interpretation of, who a certain group of people were. These texts' main concern is usually with the rulers and their entourage. The authors may touch on building sponsorship, including that of major irrigation schemes. However, questions regarding down-to-earth aspects such as technical details, building materials, maintenance, or how ordinary people made use of the engineered features remain mostly unanswered. The texts are essential, though, for information on the historical framework. The following section presents an overview of the history of Zabīd with a focus on the dynasties and events that had an impact on water use, both in the agricultural hinterland and in the city itself.[8] For a list of the important dynasties in the medieval Tihāma, see table 1.

The most significant source of information on the early history of Zabīd is the twelfth-century poet and prose writer ʿUmāra al-Yamanī's work on the history of Yemen, *Tārīkh al Yaman*.[9] The author (d. 1174) relates that Zabīd was founded in 820 by Ibn Ziyād, who had been dispatched to Yemen from Baghdad by the Abbasid caliph al-Maʾmūn (r. 813–833) to settle problems with

8 For a general outline of the history of Zabīd, see Keall 2001a, 37–40; and Sadek 2002a, 370a–371b. Chelhod (1978, 48–88) emphasizes the urban development and social history of Zabīd, while at the same time supplying political details for the history of Yemen. Smith (1987, 129–139) discusses the political history of Islamic Yemen before the first Ottoman conquest, and Stookey (1978) includes ancient South Arabia in his own work on the country's political history.

9 Most of the historical works cited here have rather flowery titles that do not reflect the subject matter of the texts. I therefore refrain from giving literal translations of the Arabic titles.

TABLE 1 Important dynasties in the Tihāma to the first Ottoman conquest

Date	Dynasty
820	Zabīd founded by Ibn Ziyād
820–1018	Ziyadids
1018–1173	Najahids, Sulayhids, and Mahdids vie for supremacy
1173–1228	Ayyubids
1228–1454	Rasulids
1454–1517	Tahirids
1516–1517	Mamluks
1517–1539	Mamluk-Lawandi military regime
1538	Ottomans arrive in Yemen
1539	Ottoman conquest of Zabīd

the population on the Red Sea coastal plain. He pitched a military camp that became the nucleus of a permanent settlement, the city of Zabīd.[10] From the tenth-century Yemeni scholar al-Hamdānī (d. mid-10th c.), who delineated the geography of his home country in his description of the Arabian Peninsula, *Kitāb Ṣifat Jazīrat al-ʿArab*, we get the impression that little occurred in the area before Zabīd was established. Al-Hamdānī mentions that Zabīd was founded in the location of the village of al-Ḥuṣayb and that the city was named after the wadi flowing by it.[11] The latter statement is repeated in more detail in the history of Yemen that was compiled by al-Janadī (d. 1332), *Kitāb al-Sulūk fī ṭabaqāt al-ʿulamāʾ wa-l-mulūk*.[12] Clearly, this indicates the crucial role of the wadi with regard to providing the economic basis for the new city.

Ibn Ziyād (r. 820–859) eventually established a dynasty in his own name. The Ziyadids ruled Zabīd and environs for the next two hundred years, and investment in infrastructure during Ziyadid rule is recorded by ʿUmāra.[13] As a city, Zabīd soon started to acquire an international reputation as a centre of learning. The tenth-century Palestinian geographer al-Muqaddasī (d. ca. 1000),

10 ʿUmāra al-Yamanī/Kay, ed. and trans. (1892) 1968, 4 (Eng. trans.), 3 (Ar. text); for information on ʿUmāra and his work, see iii–xii.

11 Al-Hamdānī/Müller, ed. 1884–1891, 1:53, lines 24–25, and 119, line 17.

12 Al-Janadī/al-Akwaʿ, ed. 1983, 221. Little is known about the author. For the available information on al-Janadī and his work, see Geddes 1965, 441a; and ʿUmāra al-Yamanī/Kay, ed. and trans. (1892) 1968, xii–xiv.

13 On the Ziyadid dynasty and its accomplishments, see ʿUmāra al-Yamanī/Kay, ed. and trans. (1892) 1968, 1–15 (Eng. trans.), 1–11 (Ar. text).

who travelled extensively in the Arabian Peninsula, describes Zabīd as one of the significant places of Yemen in his work *Aḥsan al-taqāsīm fī maʿrifat al-aqālīm*: "Al-Yaman is in two divisions, one towards the sea, low-lying, and named Tihāma. Its capital is Zabīd."[14] He states further:

> Zabīd, the capital of Tihāma, is one of its metropoles, this being the residence of the kings of al-Yaman. It is a splendid, well-built town, and popularly called "the Baghdād of al-Yaman." The inhabitants are reasonably polite, and among them are merchants, nobles, scholars, litterateurs. It is a profitable place for one who visits it, a blessed place for one who lives there. Their wells are sweet, their baths clean.... Around it are villages and cultivated fields.... Ibn Ziyād had a channel led to the town. It is an attractive town, without equal in al-Yaman.[15]

Al-Muqaddasī's reference to Baghdad, the capital of the Islamic Empire during the so-called Golden Age, is clearly to be understood as an allusion to the high level of scholarship that one would find in Zabīd.[16] His description of the villages and cultivated fields surrounding the city indicates that the Wādī Zabīd was already being controlled for irrigation farming at the time.

For almost two centuries after around 1000, rather chaotic conditions prevailed in the region of Zabīd, and various dynasties vied for supremacy—most importantly the Najahids, Sulayhids, and Mahdids.[17] Considerably more stability followed the invasion of Yemen by the Ayyubids from Egypt in 1173.[18] However, during the rather short period of Ayyubid occupation (1173–1228) there were long gaps in the governors' presence in the country, apparently due to the fact that the Ayyubids were rather uncomfortable in Yemen.[19] Therefore, one may question the extent of Ayyubid investment in Yemen, and even though Zabīd remained an administrative and economic centre, irrigation infrastructure in the Wādī Zabīd may not have been a focus of the Ayyubids.

In the final years of Ayyubid rule, a man from the eastern part of the Islamic world, most likely a native Persian-speaker, visited Zabīd. His name was Ibn

14 Al-Muqaddasī/Collins, trans. 1994, 70; see xix–xxviii on the author and his work.

15 Ibid., 82. See the Introduction to this book, section 5, on "canal" vs. "channel."

16 For a different interpretation of al-Muqaddasī's remark on Zabīd as "the Baghdād of Yemen," see Sadek 2002b, 220; and Keall 2012, 137–139. Both authors speculate that al-Muqaddasī was implying Zabīd was circular in shape, like the round city of Baghdad.

17 Smith (1987, 131–135) provides a detailed description.

18 The events leading to the Ayyubid occupation have been analyzed by Smith (1978, 31–49).

19 Smith 1969, 187 and passim; and Smith 1978, 89–90. See Smith 1987, 138–139, for a list of the Ayyubid rulers in Yemen and their dates of office.

al-Mujāwir (d. before 1250). He had travelled to Arabia to perform the pilgrimage and then continued south to Yemen and eventually proceeded along the southern coast of Arabia to the Gulf region. Probably a businessman himself, he had a keen interest in trade and commerce and the people involved in these activities in the widest sense. His account of the journey that he undertook, *Tārīkh al-mustabṣir*, contains a wealth of information on markets and the commodities found there, on prices and units of measurement, on food and agriculture, and on the social customs of the people whom he met. His work reflects a good sense of humour and enormous curiosity for anything new or unusual, including deviant social practices. It is an important source for the economic and social history of Yemen—and Zabīd in particular—under the Ayyubids.[20]

The transfer of power from the Ayyubids to their successors, the Rasulids (1228–1454), happened bloodlessly. They had arrived in Yemen with the early Ayyubids, as their army commanders, and the Rasulid rulers developed a genuine interest in their new homeland. Zabīd became their winter residence, and the Rasulids made major investments in infrastructure, including irrigation devices in the Wādī Zabīd. The authority for this time is al-Khazrajī, the official court historian of the Rasulids. He died in 1410, at the age of over seventy, having lived under four Rasulid sultans. His chronicle of the dynasty is entitled *al-ʿUqūd al-luʾluʾiyya fī akhbār al-dawla al-rasūliyya*.[21] As al-Khazrajī was a contemporary of many of the events that he reports, the information given in his work has a different quality than, for instance, ʿUmāra's references to incidents that had happened more than three hundred years before his time.[22] For our purposes, it is particularly significant that al-Khazrajī was a landowner himself and therefore familiar with, and interested in, agricultural issues. Although he does not specify where his property—which included date-palm groves—was located, one may assume that it was in the Wādī Zabīd since he was for some time professor of Koran recitation in the prestigious mosque of Mimlāḥ, a village close to Zabīd.[23]

The Rasulid rulers took great personal interest in agriculture and horticulture. New plants were introduced into Yemen. For instance, al-Khazrajī reports

---

20 Ibn al-Mujāwir/Löfgren, ed. 1951–1954; and Ibn al-Mujāwir/Smith, trans. 2008. For information on the author and his work, see Ibn al-Mujāwir/Smith, trans. 2008, 1–28.

21 Al-Khazrajī/Redhouse, trans., and ʿAsal, ed. 1906–1918. See Bosworth 1978, 1188b, for details on the author.

22 See Keall 2012, 135, for a cautionary note about relying too heavily on details in ʿUmāra's history of Yemen.

23 Al-Khazrajī/Redhouse, trans., and ʿAsal, ed. 1906–1918, 2:153–154 and 178–179 (Eng. trans.), 5:175 and 202 (Ar. text).

among the events of the year 1367 the arrival of two envoys from South Asia (Cambay and Sind) at the court of the sixth Rasulid sultan, al-Afḍal al-ʿAbbās (r. 1363–1377) who brought with them different varieties of pepper-tree plants and other rarities.[24] In the month of May 1369, the ambassadors of Abyssinia and Calicut visited the sultan's court, offering him a great variety of exotic plants, among them roses and white and yellow jasmine. The sultan gave the order to grow them in the royal garden.[25] The seventh Rasulid sultan, al-Ashraf Ismāʿīl I (r. 1377–1400), created the Garden (*bustān*) of Upper Siryāqūs near Zabīd, where he cultivated exotic trees. He is also credited with being the first to grow rice in the Wādī Zabīd, apparently with success.[26] When the sultan's youngest sons were circumcised in 1392, he hosted a splendid feast that featured a great number of fragrant flowers such as narcissi and jasmine; he also received a variety of plants as a gift from a government official.[27] Horticultural experiments were conducted in the royal gardens, and the observations and results, the failures and successes were described in texts.[28] Several agricultural manuscripts from Rasulid Yemen were actually compiled by the sultans themselves, or at least in the name of a sultan.[29]

---

24 Ibid., 2:118 (Eng. trans.), 5:135 (Ar. text).

25 Ibid., 2:119–120 (Eng. trans.), 5:139 (Ar. text).

26 Ibid., 2:270 and 287 (Eng. trans.), 5:300 and 318 (Ar. text). Another garden had been planted just outside Zabīd's city wall by the fourth Rasulid sultan, al-Mu'ayyad Dāwūd (r. 1296–1321); see ibid., 1:301–302 (Eng. trans.), 4:402 (Ar. text). Apparently, the first Rasulid sultan, al-Manṣūr ʿUmar (r. 1228–1250), had already enclosed a garden south of Zabīd, Bustān al-Manṣūriyya; see ibid., 2:11 (Eng. trans.), 5:13 (Ar. text).

27 Ibid., 2:205 and 207–209 (Eng. trans.), 5:230 and 232–234 (Ar. text).

28 The main source for descriptions of these experiments in the royal gardens is *Bughyat al-fallāḥīn*, composed around 1370 by the sixth Rasulid sultan, al-Afḍal al-ʿAbbās (r. 1363–1377); see Meyerhof 1944, 56–57, 59–62, and passim; and Serjeant 1974, 35–36 and 54. See Varisco 2002, 338–349, for the range of horticultural species that were cultivated in the Tihāma and, more specifically, along the Wādī Zabīd during Rasulid times. Many of the plant species listed are not indigenous to Yemen.

29 For an annotated bibliography of these texts, see Varisco 1989. Daniel M. Varisco has also prepared an edition, translation, and analysis of a thirteenth-century agricultural almanac that was compiled by the third Rasulid sultan, al-Ashraf ʿUmar (r. 1295–1296), as one of fifty chapters composing his major work on astronomy, *Kitāb al-Tabṣira fī ʿilm al-nujūm*; see Varisco 1994. Currently, Varisco is working on what is arguably the single most important—and earliest—agricultural treatise from Rasulid Yemen, the *Milḥ al-malāḥa fī maʿrifat al-filāḥa*, written by the same al-Ashraf ʿUmar; see Varisco 2006; Varisco 2010; and Varisco 2015, 9. In 1998, Varisco copublished (together with Smith) a mixed manuscript from the personal library of al-Afḍal al-ʿAbbās, the sixth Rasulid sultan (r. 1363–1377). The manuscript is an anthology that covers a range of subjects including—among many others—agriculture, astronomy, and medicine, in addition to archival materials from the Rasulid court. Varisco's English translation of the (anonymous) brief description of

Besides supporting agriculture, the Rasulid rulers were great patrons of other sciences, and the Rasulid era in Yemen was shaped by a number of individuals who have been described as "gifted intellectuals."[30] The Rasulids were largely responsible for making the thirteenth and fourteenth centuries one of the most prosperous periods in Yemen's history. The flourishing of Zabīd during this time is best described by the fourteenth-century traveller Ibn Baṭṭūṭa (d. 1368/9), who visited the city during the reign of the fifth Rasulid sultan, al-Mujāhid ʿAlī (r. 1321–1363), most probably around the year 1331, and described Zabīd as

> a great city in al-Yaman.... [A]fter Ṣanʿāʾ there is no [place] in al-Yaman that is larger than it nor whose population is wealthier. It lies amid luxuriant gardens with many streams and fruits, such as bananas and others, and is in the interior, not on the coast, and one of the capital cities of al-Yaman. It is a great and populous city, and contains groves of palms, orchards and running streams—[in fact] the pleasantest and most beautiful town in al-Yaman.[31]

Government interest in agriculture continued under the Tahirids (1454–1517), and many of the policies and practices of their Rasulid predecessors seem to have remained unchanged, including the use of Zabīd as their winter residence. Invaluable first-hand information for the Tahirid era is provided by the single most important historian for the city of Zabīd, a man by the name of Ibn al-Daybaʿ (d. 1537). Born in Zabīd in 1461, he spent nearly his entire life in the city, where he became a professor of *ḥadīth* (the Islamic discipline of the sayings and acts of the Prophet Muḥammad) in the Grand Mosque. Ibn al-Daybaʿ composed a history of Zabīd, which he dedicated to his chief patron, the fourth (and last) Tahirid sultan, al-Ẓāfir ʿĀmir II (r. 1489–1517). In this work, *Bughyat al-mustafīd fī akhbār madīnat Zabīd,* he describes the city and events that shaped it from its foundation by Ibn Ziyād in 820 to the year 1495. Ibn al-Daybaʿ also wrote a supplement, *al-Faḍl al-mazīd ʿalā Bughyat al-mustafīd fī akhbār madīnat Zabīd,* that brings the history of the city to 1518, the year after

---

the agricultural system in the Wādī Zabīd, compiled around 1376 and referred to as the *Taqwīm* (Varisco and Smith 1998, 277), forms the basis of Varisco 2002. Earlier, Varisco translated a royal crop register for the year 1372 from the anthology that is attributed to al-Afḍal al-ʿAbbās himself; see Varisco 1991.

30 Smith 1987, 136. As an example, see Varisco 1994, 14–16, for the wide range of scientific topics that the third Rasulid sultan, al-Ashraf ʿUmar (r. 1295–1296), addressed in his writings.

31 Ibn Baṭṭūṭa/Gibb, trans. 1962, 2:366.

al-Ẓāfir ʿĀmir II's death in the highlands at the hands of the Mamluks.[32] Ibn al-Daybaʿ is silent on the last twenty years of his life when Zabīd was under foreign occupation.

In 1516 the Egyptian Mamluk sultanate and the Ottoman Turks jointly invaded Yemen. The expedition was commanded by a Mamluk, and the Ottomans contributed a force of two thousand Lawandis. The Lawandis, a heterogeneous group of freebooters from Mediterranean coastal lands, the Levant (hence their Arabized name), were conscripted into the Ottoman navy. They were part of the invasion force that took Zabīd in June 1516 and proceeded to subdue the country from there.[33]

In the following year, 1517, the Mamluk sultanate in Cairo was overthrown by the Ottomans, and the situation in Yemen became complicated as there was now an army stranded in the country without any state authority. Between 1517 and 1539, a semi-autonomous Mamluk-Lawandi military regime ruled in Zabīd. Their dominance was confined largely to the Tihāma coastal plain. A succession of Mamluk and Lawandi military officers assumed power, living precariously and murdering one another. None of them were around long enough to make major investments, nor did they rule under conditions that would have allowed them to do so, with the exception of the last one, Iskandar Mawz (r. 1530–1536). Finally, in 1538, the Ottomans came to Yemen as their right of conquest. Zabīd was taken in 1539. This marks the end of the Mamluk-Lawandi regime. Ottoman occupation lasted for almost a century, until 1636.

The sixteenth-century historical writer al-Nahrawālī (d. 1582) provides information on these rather chaotic times and the individuals who played a part in them. Al-Nahrawālī had family ties to Yemen but spent most of his life in Mecca. Among his works is a chronicle of Yemen that he completed in 1573, *al-Barq al-yamānī fī l-fatḥ al-ʿuthmānī*.[34] It covers the history of the country from the beginning of the sixteenth century through to the years of the early Ottoman occupation. Al-Nahrawālī's writing is, not surprisingly, pro-Ottoman in tone. After all, he was well connected with the ruling Ottoman elite, and his history of Yemen was commissioned by a high-ranking Ottoman official.

---

32 For both volumes, see Ibn al-Daybaʿ/Chelhod, ed. 1983. General information on Ibn al-Daybaʿ and his work is given by van Arendonk (1968, 746a–b) and, most recently, by Kenderova (2014, 178–183).

33 See Blackburn 1971, 31–65, for details of the Mamluk-Lawandi conquest and the period of their occupation.

34 Al-Nahrawālī/al-Jāsir, ed. 1967. Detailed information on the author can be found in Blackburn 2005, xi–xvi.

2 The Water Resources

Quite naturally, looking at the reality of Zabīd's geographical location makes one wonder how the city could not only survive but thrive under the local climatic conditions. Zabīd lies midway across the Red Sea coastal plain, the Tihāma, one of the hottest areas of the Earth, and receives an average annual rainfall of less than 150 mm.[35] This figure reflects a long-term average, though, and the annual variability of precipitation is high. In exceptional years, direct rainfall can be sufficient to produce a dry-farmed crop, and farmers sow millet seeds during the days immediately following what they judge to be adequate precipitation to sustain the crop to maturity. But such a rainfall pattern is unusual and cannot be relied upon; at best, it provides for only a very few subsistence crops. Viable agriculture in the Wādī Zabīd has always depended on irrigation.

The city of Zabīd is located at the intersection where a minor side wadi, the Wādī Sanha, joins the main flood course of the Wādī Zabīd, one of the major wadis of western Yemen (see fig. 44); the Wādī Zabīd drains a mountainous catchment area, flowing in a westerly direction towards the Red Sea. The biannual *sayl* in the wadi results from the pattern of rains in the mountains during spring and late summer that were described in the Introduction to this book.[36] The *Taqwīm*, an anonymous summary of the agricultural system in the Wādī Zabīd from the upper to the lower parts that was compiled around 1376,[37] gives us detailed information on the anticipated times of the flood seasons in the wadi. As the text states, the spring *sayl* can be expected at the end of April, and the stronger summer floods typically happen during the month of August and the beginning of September but may start as early as July. While a weak flood can occur in the interlude before the next spring, this is judged to be an exception.[38] After the summer spate recedes, only the base flow (*ghayl*) remains in the upper reaches of the wadi.[39]

---

35 Kopp 1981, 42, table 2, and 44, figs. 5a and b; and Remmele 1989, 31, fig. 4, and 32, fig. 5.

36 Al-Eryani (1979, 43, 50) gives a median annual rainfall of 665 mm for the Wādī Zabīd catchment area, the size of which he specifies as 4340 km^2.

37 See n. 29 above.

38 Varisco 2002, 324 and 326–328. Varisco also cites an unpublished development-agency report that lists the number of flood days recorded for the individual months of the year 1970 as follows: April, 5; May, 2; June, 3; July, 16; August, 16; September, 8; October, 4 (325, n. 5). Please note that the dates listed in the *Taqwīm* are given according to the Julian calendar. Since the text was compiled around 1376 and thus well before the Gregorian calendar reform in 1582, the dates are some eight days earlier than in our modern calendar; see Varisco 1991, 6.

39 Varisco (2002, 328) uses the term "spring flow," which is the literal translation of *ʿayn*, the word that is indeed used in the Arabic text; see Varisco and Smith 1998, 277, line 28.

Until fairly recently, before diesel-operated well pumps were introduced on a large scale, the spate in the Wādī Zabīd provided the main source of water for the city and its agricultural hinterland, both for irrigation through diversion of the *sayl* and for domestic use through wells tapping the groundwater that was replenished by the biannual spate. In this context, it is important to note that today when locals speak of "the Wādī Zabīd," they are referring to the land watered by the wadi, not just the flood course itself. This emphasizes that in their understanding the two form one integrated system: without water, the land has no value, and without land, the water cannot be made useful.[40]

I would like to suggest, though, that "base flow" is more appropriate in the context of a wadi: when the *sayl* (which is mostly the result of direct surface runoff) dries up following the end of the rains in the catchment basin, a (gradually decreasing) stream flow in the upper reaches of a wadi may be sustained in between rainy seasons, resulting from groundwater seepage. Technically, it is called "base flow" or "drought flow." This is what locals in the Wādī Zabīd—and in other wadis of Yemen—today refer to as *ghayl*. See also Kopp 1981, 50 and 52; and Bonnenfant 1997, 18. In Yemeni dialect, *ghayl* has the general meaning of a perennial stream—which may, of course, be fed by a spring. But this is only one option, and there are others; see Part III, Introduction and Case Study 6.

40 The same observation is reported for the Tihāma by Escher (1976, 61) and Donaldson (2000, 134).

CASE STUDY 2

Sayl Irrigation in the Wādī Zabīd

In this chapter I will address the technical side of the *sayl*-irrigation system in the Wādī Zabīd before the first Ottoman conquest in 1539. The individual components and their function are described, as well as their terminology. I will then discuss the management of the seasonal water resources, in particular the water-allocation rights that allowed the system to operate generally smoothly. Physical remains and historical sources, most importantly texts from Rasulid times, provide the details. However, it is the study of the traditional irrigation system in present-day Wādī Zabīd that contributes much to our understanding of both the engineering and the organizational side of the medieval system.[1] Therefore, the chapter starts with a brief account of the traditional system, which will be reviewed further in the context of the water-allocation rights. The overall picture that emerges is characterized by continuity and community-based engagement.

1 The Traditional *Sayl*-Irrigation System and its Terminology

The following description of the traditional irrigation system in the Wādī Zabīd is based on fieldwork conducted since the early 1990s. The terminology is that used by today's farmers. A small number of twentieth-century studies focus on irrigation in the Wādī Zabīd and were employed to cross-check the terminology; they are listed below. Needless to say, irrigation terminology is never unambiguous, and even within the Wādī Zabīd, we can observe some variations in the use of certain terms. Works that do not directly refer to the Wādī Zabīd are not included here since the terminology in other parts of Yemen can be quite different.

Three studies are of significance in this context. First, the 1971 "Survey of the agricultural potential of the Wadi Zabid, Yemen Arab Republic. Land tenure and water rights," a first-hand account written for the development agency Tesco-Viziterv-Vituki, provides essential information on the traditional

1 See also Varisco (2015, 96–101) who points out the value of anthropological observations in today's Yemen for making sense of medieval Arabic manuscripts "because there is an obvious continuity in the overall trajectory of Yemen's agricultural history" (97).

 | DOI:10.1163/9789004387713_006

FIGURE 29 Simple diversion barrage for irrigation of fields in the low-lying areas along the flood course of the Wādī Zabīd

system in operation before the implementation of the Tihāma Development Project (completed in 1979). Second, Robert B. Serjeant's 1988 "Observations on irrigation in south west Arabia" gives a good general overview of traditional *sayl*-irrigation practices in the Tihāma, with specific reference to the Wādī Zabīd, though the article does suffer from some inconsistencies. Third, Paul Bonnenfant's 1997 publication, "La maîtrise de l'eau dans le wadi Zabid, Yémen," describes the *sayl*-irrigation system after the implementation of the Tihāma Development Project.

Today in the Wādī Zabīd, one can observe parts of the traditional system of water diversion and distribution operating side by side with the permanent barrages and primary canals built of concrete in the late 1970s as part of the Tihāma Development Project. Simple barrages consisting of heaped-up earth and coarse gravel are often reinforced with brushwood, palm fronds, and boulders. They reach into the flood course at an acute angle in relation to the flow direction. They divert particle-rich water for the purpose of sedimentation and land reclamation along the sides of the wadi bed. Even today, the fields created in the low-lying areas along both banks of the wadi receive water in this way (fig. 29), a supplement for the farmers that is independent of, and in addition to, the water provided from the main irrigation structures. In fact, a specific term is used in the Wādī Zabīd for such a field: it is called a *jalla*, pl. *jalal*.

The small earth-and-gravel barrage diverting water from the wadi bed onto it is a *manṣūb*, pl. *manāṣīb*[2] (an alternative term is *muḥammal*).

Depending on the force of the spate, such traditional diversion barrages are eventually washed away during the *sayl* (fig. 30) and have to be rebuilt before the next rainy season. For the farmers in the Wādī Zabīd, their reconstruction is nothing unusual. It forms part and parcel of the work during the agricultural year (see fig. 31).

A larger diversion barrage, which is also referred to as a *manṣūb* or *muḥammal*, diverts the *sayl* into a canal that leads towards the fields lying above the banks of the wadi.[3] In local dialect, the term *sharīj*, pl. *shuruj*, denotes the canal that is connected to a diversion barrage at its head—that is, a primary canal.[4] *Sharīj* also implies the system of contiguous fields irrigated by it. This linguistic detail underlines once more the fact that the land and the water form a unit. A smaller secondary canal is referred to as a *sāqiya*, pl. *sawāqī*.[5]

Within a system of contiguous fields, the water is distributed along the irrigation canals following the principle of gravity flow. A small temporary dike of mud and brushwood (*maʿqam*, pl. *maʿāqim*) is built across a canal to direct the water into the adjoining field.[6] This field is submerged first, after which its field bank (*zabīr*, pl. *zubur*)[7] is breached and the water flows into the next field in succession, and so forth (field-to-field irrigation). Such a breach in a *zabīr* is called a *kharāb*. No field receives water until irrigation of the one preceding it in the cycle has been completed. The water is therefore distributed in a fixed order to the entire unit of fields irrigated by a specific *maʿqam*. As with the term *sharīj*, a *maʿqam* also refers to this unit of fields, and it constitutes a subdivision of a *sharīj*. Eventually, the small dike is pulled down and another one erected slightly farther along the canal in order to start irrigating the next unit of fields. The field banks enclosing the modern fields are up to 1 m high to allow irrigation by deep ponding (fig. 32). A field contained by such *zubur* is called a

---

2 Both terms, *jalla* and *manṣūb*, are mentioned in Tesco-Viziterv-Vituki 1971, 13; for *manṣūb*, see also Bonnenfant 1997, 21.

3 Some sources cite the terms *ʿaqm*, pl. *ʿuqūm*, or *maʿqam*, pl. *maʿāqim*, for a diversion barrage erected across the wadi; see Serjeant 1988, 145, 148, and 149; and Bonnenfant 1997, 15. See section 2 below for *ʿaqm* and *maʿqam* with the meaning of a diversion barrage that diverts part of the *sayl* into a primary canal (*sharīj*).

4 Tesco-Viziterv-Vituki 1971, 14, mentions the term *sharīj* with the general meaning "canal"; for "primary canal," see Bonnenfant 1997, 14.

5 Bonnenfant 1997, 14, n. 11.

6 Tesco-Viziterv-Vituki 1971, 14. See also n. 3 above.

7 Tesco-Viziterv-Vituki 1971, 25; Serjeant 1988, 148; and Bonnenfant 1997, 14.

FIGURE 30 The Wādī Zabīd during a seasonal spate, with a traditional diversion barrage partly washed away

FIGURE 31 Farmer heaping up a diversion barrage in the Wādī Zabīd

FIGURE 32 Field enclosed by high field bank to allow irrigation by deep ponding

zahab, pl. *zāhīb*.[8] By contrast, the term used for a field that is fed by occasional rainfall, not irrigated from the wadi, is *jirba*, pl. *jirab*. A *jirba* does not require such high field banks.

Since the volume of the spate varies from year to year, the fields closer to the wadi course—that is, closer to the intake of the primary canal—are irrigated first, and the areas towards the farther reaches of the farmed land receive water only in those years when there is a sufficiently strong *sayl*. This practice is dictated by the requirement to use the limited water resources efficiently. Therefore, in addition to the natural overall gradient of the wadi, which flows more or less east-west, the entire irrigated land of the Wādī Zabīd is roughly terraced, stepping down from the highest points adjoining the wadi course in the middle towards the lowest reaches along the northern and the southern fringes. A "terrace" in this context means a *sharīj* system of contiguous fields with water distribution from field to field.

At the end of a terrace, either between fields or built into the course of a canal, engineered devices must be provided to accommodate drops in elevation. These structures are best described as "spillways" (*jusūr*, sg. *jisr*).[9] They provide a safe way to release excess water from a terrace into a discharge area by allowing a controlled overflow and preventing cut-back erosion. To maintain their effectiveness, the height of the spillways has to be raised at regular in-

8 Tesco-Viziterv-Vituki 1971, xiv; and Bonnenfant 1997, 15.

9 Bonnenfant 1997, 14–15.

FIGURE 33 Spillway in the Wādī Zabīd

tervals to keep pace with the rising level of the fields due to sedimentation. In the past, the spillways were invariably built with baked bricks and lime mortar (fig. 33). Today, there is a tendency to use concrete blocks and Portland cement, but the principle remains the same.[10] In no case has it been possible to determine the date of the earliest phase of a spillway, but as many as six rebuildings up to the present have been recorded. The spillways reflect continuity in water management, employing a simple but effective device to create stability in an otherwise volatile landscape.

The present-day farmers of the Wādī Zabīd are familiar with the problem of the continually rising level of the fields. They solve it by extending the diversion barrages farther upstream, thereby gaining additional height to operate the system using gravity flow.

2 Physical Remains of the *Sayl*-Irrigation System and its Terminology in Pre-1539 Historical Sources

Barrages can be seen exposed in section in high scarps carved at the sides of the wadi by flash floods. These were originally built of earth, coarse gravel, and boulders, all taken from the flood course, and set on what was then the bare

10 The modern version is called a *manfadh*. See also Serjeant 1988, 148.

FIGURE 34 Section of archaeological remains of a diversion barrage built of earth, coarse gravel, and boulders buried beneath sediments at the side of the Wādī Zabīd (1 m scale)

wadi bed (fig. 34). They represent early attempts to divert the spate, with the aim of trapping particle-rich water and creating a field from the settled sediments. Close examination reveals how the sediments built up against these barrages and rose steadily over time, commensurate with the process described at Mārib (see Case Study 1). Eventually, the barrages were completely buried, and the field was enclosed by perimeter banks using the deposits that had accumulated in the meantime. Bigger and higher diversion structures then needed to be projected into the wadi farther upstream. They had to be connected to an increasingly long canal network to distribute the water as the area of irrigated land kept extending.

Traces of such barrages have been found in different reaches of the Wādī Zabīd. This is because every so often a particularly strong *sayl* would cut into the rising fields (even today the farmers of the Wādī Zabīd complain about the shifting flow behaviour of the *sayl*), resulting in the exposure of a scarp, which allows the sediment accumulations to be seen in section, with the barrage at the bottom, level with the wadi bed (fig. 34). Above it, more recent field banks with heights of up to 1 m can be observed in the scarp.

As yet, it is not known how early this practice of wasteland reclamation along the sides of the wadi bed began. With respect to the Tihāma, a distinction between a *zahab* (a field contained by banks to allow irrigation from the *sayl* by deep ponding) and a *jirba* (a field without banks because it is fed by occasional rainfall) was already being made by the tenth-century writer al-Hamdānī.[11] A *zahab* requires planning and investment.

The maximum height of field sediments that has been measured along the sides of the wadi course is 7 m, reflecting centuries of controlling the spate and using it for irrigation. It is interesting to observe in section that in these sediments the content of coarse particles, such as sand and grit, is low. This is clear evidence of good operation and maintenance of the *sayl*-diversion system insofar as it shows that only the fine particles reached the irrigated fields while the coarse ones were caught by the irrigation structures farther up. The occurrence of thick sand layers in the fields would constitute a serious threat to irrigated agriculture, mostly because of the low capacity of sand (and grit) to retain water and nutrients. It would also indicate that the delicate balance of control was undermined by management deficiencies.[12]

We know from historical sources that the system of *sayl* control and irrigation in the Wādī Zabīd received major attention during Rasulid times. So far, the most detailed evidence can be found in a tax register of the fourth

11 Al-Hamdānī/Müller, ed. 1884–1891, 1:199, lines 12–20. Müller collated four different manuscripts of al-Hamdānī's work and used a fifth one for additional annotations (2:v–ix). In the edited Arabic text, the term *dhahab* is cited instead of *zahab* (1:199, line 20). However, the two manuscripts in the Staatsbibliothek Berlin, Orientabteilung, that Müller had used for preparing the text edition indicate the use of *zahab*. In the 1405 copy, Mq. 730, 273, line 10, *z-h-b* is clearly legible, including the diacritical points, even though the manuscript is otherwise mostly lacking in diacritics. In the 1658 copy, Mf. 1196, 133, line 12, *z-h-b* is also clearly legible, including the diacritical points, and the vocalization is specified as *zahb*. In his annotations, Müller had already pointed out that in the manuscript Codex Huber, Strassburger Landes—und Universitätsbibliothek, the root *z-h-b* is used (2:211). I confirmed this in the Bibliothèque nationale et universitaire Strasbourg, MS 4268, 144r, line 6 (vocalized *zihb*). Müller's misreading of *dhahab* for *zahab* in both manuscripts housed in Berlin and his preference for *dhahab* over *zahab* in case of the Strasbourg manuscript may be related to the fact that *z-h-b* is not a root that is found in classical Arabic. For instance, it is not listed in Lane's *Arabic-English Lexicon*. See Serjeant 1974, 34–35; and Varisco 2009, 386–387, for English translations of the passage based on Müller's text edition. For general information on the 1405 and 1658 manuscripts, see al-Hamdānī/Müller, ed. 1884–1891, 2:viii–ix, "B" and "E"; and Ahlwardt 1893, 5:382–384 (nos. 6059 and 6060). For general information on the manuscript Codex Huber, copied in 1856, see al-Hamdānī/Müller, ed. 1884–1891, 2:ix, "D"; and Kusaybī 1985, 54–55 (no. 51).

12 Unfortunately, large quantities of coarse deposits create a problem in the contemporary canal system in the Wādī Zabīd, as described by Serjeant (1988, 146–147) and confirmed during my fieldwork. This makes transport of the coarse particles farther into the fields easier and at the same time reduces the canal capacity.

Rasulid sultan, al-Mu'ayyad Dāwūd (r. 1296–1321).[13] It was compiled during the first years of his reign—that is, at the very end of the thirteenth century.[14] This manuscript is subdivided according to the Rasulid administrative regions, with a substantial section on Zabīd and its agricultural hinterland, including a unique map of the Wādī Zabīd (fig. 35). Unlike in ancient Mārib, no diversion barrage was built across the mouth of the gorge where the wadi debouches from the mountains because the valley of the Wādī Zabīd is wider and not as easily closed off. Instead, the map shows a succession of individual main canals (*shuruj*) along the central flood course of the wadi that leads from the mountains in the east (at the top of fig. 35) to the Red Sea in the west (at the bottom), in modern terms a distance of some 45 km. The city of Zabīd is located on the northern bank. Five canals divert the water onto the southern bank of the wadi, and twelve onto the northern one.[15] The tax register also contains a list of irrigation areas (*shuruj*) in the Wādī Zabīd and the levies derived from each of them.[16] For an inventory of the names, see table 4 (second column from right). Clearly, the irrigation system was firmly established by this time.

From the manuscript we also learn about the Rasulids' investment in irrigation infrastructure in the Wādī Zabīd. For instance, Maʿqam al-Bunay, the highest-lying diversion barrage in the wadi, was particularly vulnerable to the force of the *sayl* and was regularly damaged or even washed away. Reconstruction was paid for by the government—the farmers were not expected to contribute to the costs.[17] This shows that it was understood that Maʿqam al-Bunay needed to be in place not only to allow irrigation in Sharīj al-Bunay but also to break the initial force of the *sayl* and thus protect the infrastructure farther downstream. The term *maʿqam* (and *ʿaqm* as a synonym) is used in the manuscript for the diversion barrage at the head of a primary canal (*sharīj*) that diverts part of the *sayl* into it.[18] The tax register distinguishes between the upper, the middle, and the lower part of the canal system in the wadi, and this division (*al-shuruj al-ʿulyā, al-shuruj al-wusṭā*, and *al-shuruj al-suflā*)

13 The manuscript was edited by Jāzim (2008). While one might expect to find detailed information on the irrigation system in the agricultural almanacs from the Rasulid era, this is not the case. The almanacs focus on crops and their varieties including sowing and harvesting times but usually mention irrigation only in general terms.

14 Vallet 2010, 74–75.

15 Jāzim 2008, 387 (facsimile of the map); see 18 for a schematic drawing. The original version of the map uses diacritical signs rather sparingly (387), while the editor has taken certain liberties in adding these on the schematic drawing (18). They should be used with caution.

16 Ibid., 41–44.

17 Ibid., 11.

18 Ibid., 10.

FIGURE 35 Map of the Wādī Zabīd from the tax register of the fourth Rasulid sultan, al-Mu'ayyad Dāwūd (r. 1296–1321)

is repeated in the *Taqwīm* of 1376[19] and confirmed by the fourteenth-century historian al-Khazrajī.[20]

A primary canal by the name of Sharīj al-Marḍī is mentioned by Ibn al-Dayba‘ in his chapter on the reign of the second Rasulid sultan, al-Muẓaffar Yūsuf (r. 1250–1295).[21] According to al-Khazrajī, the seventh Rasulid sultan, al-Ashraf Ismā‘īl I (r. 1377–1400), had estates in Sharīj al-Minqāz that he inspected in the spring of 1395.[22] A Sharīj Abīra is referred to in a list of incidents in 1396,[23] while seven primary canals of the upper wadi reaches are listed by name for the year 1383;[24] see table 4 (right column).[25]

Al-Khazrajī also reports the breaking of a barrage (*‘aqm*) in the Wādī Zabīd by a strong *sayl* in 1399.[26] He refers to "al-‘aqm al-kabīr al-Mujāhidī." This implies that the barrage was erected by al-Mujāhid ‘Alī, the fifth Rasulid sultan (r. 1321–1363). Al-Khazrajī's phrasing—"al-‘aqm al-kabīr" ("the great barrage")—suggests that the structure was more solid and intended to be more permanent than a traditional diversion barrage built of earth and coarse wadi gravel. Or was the breaking of "the great barrage" recorded for posterity because al-Khazrajī's job as the court historian was to refer to the greatness of the Rasulid rulers and their works wherever possible?

In any case, the historical sources provide firm evidence of an engineered system of spate control and irrigation in the Wādī Zabīd as early as the thirteenth century, and we may safely suppose that its principles were established well before that time.

3 Water-Allocation Rights in the Wādī Zabīd

Studying a 1973 aerial photograph of the Wādī Zabīd, Horst Kopp noticed the rather irregular shape of the fields. He describes the clear signs of wear: original

19 See Introduction to Part 2, n. 29. In his English translation, Varisco (2002, 327) reads "al-sharj al-‘ulyā" and translates as "the upper part of the wadi." Instead, one should read "shuruj" and translate as "the upper canals" or "the land irrigated by the upper canals." See Varisco and Smith 1998, 277, lines 10 and 32–34, for the Arabic text.

20 Al-Khazrajī/Redhouse, trans., and ‘Asal, ed. 1906–1918, 2:55–56 (Eng. trans.), 5:68 (Ar. text).

21 Ibn al-Dayba‘/Chelhod, ed. 1983, 91.

22 Al-Khazrajī/Redhouse, trans., and ‘Asal, ed. 1906–1918, 2:243 (Eng. trans.), 5:272 (Ar. text "Minqār").

23 Ibid., 2:253 (Eng. trans.), 5:284 (Ar. text).

24 Ibid., 2:153 (Eng. trans.), 5:174 (Ar. text).

25 Interestingly, two of the canals (al-Māwī and al-Badānī/al-‘Ibrī) are classified as belonging to the middle reaches of the Wādī Zabīd before implementation of the Tihāma Development Project in the late 1970s.

26 Al-Khazrajī/Redhouse, trans., and ‘Asal, ed. 1906–1918, 2:284 (Eng. trans.), 5:314 (Ar. text). See n. 3 above for the term *‘aqm*.

right angles had become rounded and straight lines had become crooked over time as sediment deposits and successive repairs obscured the pristine layout. By comparison with the more rectangular pattern of the fields in other wadis in the Tihāma, he concludes that the irrigation system in the Wādī Zabīd was put in place earlier and had been functioning in more or less the same way for a considerably longer period than elsewhere.[27] Kopp speculates that the reason for the exceptional continuity in the Wādī Zabīd was the application of water-allocation rights that had survived for several hundred years. Before the question of their origin is discussed, it is necessary to address some general considerations regarding water-allocation rights in Islam.

3.1 *Water Law and Water-Allocation Rights in Islam*

The ordering principles, rules, and regulations that govern all aspects of a Muslim's existence constitute the code of Islamic law (*sharīʿa*).[28] The first source of the *sharīʿa* is the Koran, which according to Muslim belief is God's word as revealed to the Prophet Muḥammad and eventually compiled and written down in the form of a book. While the Koran provides more general guidelines, it does not give answers to all the specific questions and problems that emerge in day-to-day situations. Therefore, the sayings and exemplary ways of acting attributed to Muḥammad (his *Sunna*) became the second source of the *sharīʿa*. A *sunna*, pl. *sunan*, of the Prophet is communicated in a *ḥadīth*, pl. *aḥādīth* (literally, a "report"). As with the Koran, the *aḥādīth* were gathered and are preserved and transmitted in literary form, the *Ḥadīth* (corpus of *aḥādīth*). The science of law, or jurisprudence (*fiqh*), developed in two distinct ways for the emerging main branches of the new religion, Sunni and Shiite Islam. Both refer to the Koran and the Prophet's *Sunna* as primary sources, though different *ḥadīth* collections are used. Theological and juristic disagreements eventually resulted in the recognition of three schools of law in Shiite Islam, and four in Sunni Islam.[29]

With regard to Islamic water law, in the absence of specific Koranic legal provisions, it is the Prophet Muḥammad's *Sunna* that provides the framework for adjudication. The Prophet is reported to have said that "Muslims have common share in three [things]: grass, water and fire."[30] Water in its natural state is

---

27 Kopp 1981, 128 and plates 2 and 3.

28 Interestingly, the literal meaning of *sharīʿa* is "a place of descent to water" or "a way to water." See Lane 1863–1893, s.v. *sh-r-ʿ*, *sharīʿa*.

29 The formation of the different branches and schools of Islamic law is a complex matter. Kamali (2008, chapters 2–4) gives an excellent overview.

30 Abū Dāwūd/ʿAbd al-Ḥamīd, ed. 1950, 3:377 (no. 3477). The *Sunan Abī Dāwūd*, gathered by Abū Dāwūd al-Sijistānī (d. 889), is recognized by Sunnis as one of the six canonical collections of *aḥādīth*.

mubāḥ; that is, nobody owns it because it cannot be owned.[31] While the ideal of water as a public commodity and the right of free access to it is certainly appealing, the reality of life in an arid climate where water is a limited resource, as in major parts of the Islamic world, called for regulations in order to allow a community to enjoy stability and prosperity. The *shāfiʿī* school of law in Sunni Islam, named after its founder Muḥammad ibn Idrīs al-Shāfiʿī (d. 819), is authoritative for determining water law in large parts of Yemen, including Zabīd, which has been a centre of *shāfiʿī* law since Rasulid times. The following comments are therefore restricted to *shāfiʿī* law.[32]

Al-Māwardī (d. 1058), a distinguished scholar of *shāfiʿī* law, classified river water into three categories:[33]

a) Water of great natural rivers—such as the Tigris and the Euphrates—flows abundantly and can satisfy all demands, including irrigation and watering animals. Everyone has free access to it.

b) Small natural rivers are of two kinds, with the first carrying sufficient water for all the people who live in its vicinity. They share the water, and nobody is denied access. The second does not have enough water to meet the needs of everybody in its region. Here, engineered devices (barrages) are required to control the flow, and the water is allocated according to upstream priority; that is, the higher-lying areas are irrigated first and then, successively, those farther downstream. This practice has its origins in a *sunna* of the Prophet Muḥammad.

c) Water in an artificial, man-made canal is also of two kinds that determine its distribution. First, the volume of the water may be adequate to accommodate the demand of all living in its vicinity, in which case they share it. If, however, the volume is not sufficient, then only those who participated in digging the canal have the right to a share.

Clearly, in those examples where water is the limiting factor, the fundamental question is how much water an individual farmer is entitled to. The general principle is that after a field is submerged to the height of a man's ankle, its field bank should be breached to allow the water to flow into the next plot in succession, and so forth. This regulation is attributed to the Prophet Muḥammad. However, al-Māwardī explains that it is not necessarily binding for everybody

31 Maktari 1971, 13–14.

32 The best overview of the topic is given by Maktari (1971). A frequently cited general survey on Islamic water law is Caponera's *Water Laws in Moslim Countries* (1973–1978); an earlier edition was published in 1954. However, see Maktari 1971, 7, for a cautionary note on the value of this work with regard to specific legal cases in Yemen. Varisco (1983, 369, n. 6) raises serious concerns about inaccuracies contained in the book.

33 See al-Māwardī/Yate, trans. 1996, 256–259; 259–262 on the classification and sharing of well and spring water. See also Maktari 1971, 14–15 and 28.

at all times and in all places. He points out that crop requirements differ, and so do soil, climate, and the availability of water. While he and other *shāfiʿī* jurists do not consider their view as implying rejection of an explicit rule given in the *Sunna*, they interpret the regulation about ankle-deep irrigation as reflecting the custom of the specific location where the Prophet Muḥammad settled a dispute over irrigation practices.[34]

We can observe an interesting issue here that has to do with the fact that irrigation agriculture did not start with the rise of Islam but was well established in pre-Islamic Arabia. The Mārib oasis (see Case Study 1) is only one example of many, and statements that question the existence of a fully developed system of water-allocation rights in ancient times are unfounded.[35] The jurists of the early Islamic centuries encountered existing customary rules of conduct that had been shaped over many centuries in accordance with a location's specific natural conditions and in response to the requirements of a community. This situation was clearly taken into consideration by al-Māwardī and his colleagues, even though *shāfiʿī* law was originally opposed to recognizing custom as a source of law.[36] In addition, with regard to Yemen, one needs to bear in mind that the country never formed a core part of the Islamic Empire and remained largely outside its central authority, and this includes legal decision making. It is therefore not surprising that we find the strong presence of customary law (*ʿurf*) in Yemen and its eventual acceptance by the *shāfiʿī* jurists.[37] This is particularly true of water law.

As an example, we may look at a modification of the aforementioned rule specified in the *sharīʿa* and established by the Prophet Muḥammad that a field is entitled to a share of water until it is submerged to the height of a man's ankle. As described earlier, the height of the field banks in the Wādī Zabīd—both in terms of the physical remains from the past that are preserved in sediment profiles and current usage—show that here the fields were (and still are) submerged to a considerably greater depth. Clearly, the customs of the Wādī Zabīd differ from those of the location where the principle of ankle-deep

34 Al-Māwardī/Yate, trans. 1996, 257. Quite naturally, not everybody in the *shāfiʿī* school shared this view; see Maktari 1971, 34.

35 See Caponera 2001, 94, for an example: "Before the Prophet Muhammad, in the *djahiliyya* or 'period of ignorance,' water regulations were not established in Arabia." It is foreigners, though, who tend to ignore Yemen's ancient, pre-Islamic past and its implications for today's practices, while Yemenis themselves are very much aware of them. See Case Study 1, section 6.1, for rules regulating water allocation in ancient Yemen.

36 Dien 2004, 58–60; and Libson 2003, 68–70. A detailed overview of the relationship between customary law and *sharīʿa* can be found in Donaldson 2000, 42–51.

37 Maktari 1971, 29–32. In fact, most of the legal rulings presented in Maktari's book substantiate this point. Haddash (2001) outlines the important role that customary law plays in water legislation in today's Yemen.

irrigation was established. However, one underlying principle is always the same: Islamic water-allocation rights are tied to the land and cannot be separated from it.[38] Muslims understand water and land as forming one integrated system, a natural unit, and this rule is reflected in linguistic nuances, as in the examples of wadi, *sharīj*, and *maʿqam* cited above.

3.2 *Traditional Water-Allocation Rights in the Wādī Zabīd*

The technical term in Arabic for the aforementioned *sharīʿa* principle of water allocation by upstream priority is *al-aʿlā fa-l-aʿlā*, literally meaning "the one who is highest [with his plot], then the one who is next highest";[39] that is, the *sayl* is diverted first to the high-lying areas in the wadi, and then, successively, to those farther downstream. Overall, reliability of spate irrigation is therefore "a function of proximity to the wadi, and distance downstream from it."[40] Given the seasonality and limited duration of the *sayl*, upstream priority is the most efficient way to distribute the water. However, it is obvious that the landowners of the downstream areas are at a disadvantage, in particular in those years when the *sayl* is weak. At the same time, upstream priority implies that a system of diversion barrages be in place that requires repair and maintenance on a regular basis. Conflicts among the farmers are thus likely to happen.

The Wādī Zabīd is unique among the wadi systems in Yemen insofar as its water-allocation rights deviate from the *sharīʿa* principle of unrestricted upstream priority. Instead, the water law integrates principles of customary law (*ʿurf*) and allocates the water by combining upstream priority and prescribed calendar dates. In order to understand the implications, a brief summary will be given here of the traditional practices as documented at the beginning of the 1970s—that is, before implementation of the Tihāma Development Project in the late 1970s. It is based on an unpublished development-agency report, the only source that deals directly with the topic.[41] The basic parameters are valid to this day.[42]

---

38 Maktari 1971, 9 and 134.

39 Salameh 1999, 138.

40 Tihama Development Authority, n.d., 31.

41 See Tesco-Viziterv-Vituki 1971, 5–13. Kopp (1981, 125–128) recapitulates the main points. Please note that Tesco-Viziterv-Vituki reproduces the canal names without proper transliteration. Rendering the names in English for this book was accomplished with the kind assistance of two local informants, ʿAbd al-Karīm Aḥmad al-Sālimī, director of the Zabīd Office of the General Organization of Antiquities and Museums (GOAM), and Aḥmad ʿĪsā Aḥmad Hulaybī, schoolteacher of Maḥall al-Shaykh, a village in the Wādī Zabīd.

42 See Bonnenfant 1997, 16–20, for a description of the practices during the 1990s—that is, after implementation of the Tihāma Development Project. It bears close resemblance to the earlier model.

The traditional spate-irrigation system is described as comprising seventeen primary canals, each with its own intake diverting part of the *sayl*, and the areas irrigated by them. These canals are classified in three groups, an upper, middle, and lower one (see table 2).

TABLE 2 Traditional water allocation in the Wādī Zabīd before implementation of the Tihāma Development Project

	Irrigated area (ha)	**Irrigated area (%)**	**Name of canal**	**Dates of allocation rights**
Group I (upper reaches)	4,775	28	1. al-Bunay 2. al-Barrī 3. al-Jirba 4. al-Manṣūrī 5. al-Rayyān 6. al-Baqar	a) Oct 19–Mar 28: calendar date allocation of base flow (see table 3) b) Mar 29–Aug 2: upstream priority (slightly modified)
Group II (middle reaches)	10,175	60	1. al-Māwī 2. al-Yūsufī 3. al-ʿIbrī 4. al-Jarhazī 5. al-Jurayb 6. Bīra 7. al-Nāṣirī	Aug 3–Sep 13: upstream priority (slightly modified)
Group III (lower reaches)	1,450	9	1. al-Sharʿabī 2. al-Maḥraqī 3. al-Ḥaram 4. Wādī al-ʿAyn[a]	Sep 14–Oct 18: upstream priority (modified)
Fields created through land reclamation in the low-lying areas along both sides of the wadi bed	530	3		any time (usually during low floods)

SOURCE: TESCO-VIZITERV-VITUKI 1971, 5–13

a Wādī al-ʿAyn is the name of the Wādī Zabīd beyond the city of Zabīd, theoretically flowing all the way to the Red Sea.

TABLE 3 Number of irrigation days of group I canals in the Wādī Zabīd during the months of base-flow allocation (Oct 19–Mar 28), prior to implementation of the Tihāma Development Project

Canal	Oct 19– Nov 13	Nov 14– Dec 13	Dec 14– Jan 13	Jan 14– Feb 13	Feb 14– Mar 13	Mar 14– Mar 28	Total number of days
1. al-Bunay	12	—	15	16	18	5	66
2. al-Barrī	3	2	4	—	2	—	11
3. al-Jirba	4	6	3	4	3½	—	20½
4. al-Manṣūrī	7	6	9	—	4½	—	26½
5. al-Rayyān and 6. al-Baqar	—	16	—	11	—	10	37

SOURCE: TESCO-VIZITERV-VITUKI 1971, 7 AND 9
Note: The sizes of the cultivated areas are as follows:
al-Bunay and al-Barrī: 1,250 ha
al-Jirba: 600 ha
al-Manṣūrī: 1,200 ha
al-Rayyān: 1,250 ha
al-Baqar: 475 ha

Group I consists of six individual canals (listed sequentially from higher-lying to lower-lying intakes) that irrigate the fields in the upper reaches of the wadi. They have the right to use the water October 19–August 2. This may appear a long period. However, it is unlikely that there will be a *sayl* October 19–March 28. Therefore, only the base flow is available for irrigation. It is allocated according to prescribed calendar dates; see table 3 for the exact schedule.[43] The second part of the term includes (in many years) the time of the spring spates. The sequence of water sharing follows upstream priority, with the exception of canal no. 4, which has priority over no. 3, and canals nos. 5 and 6 (the lowest ones of this group), which are operated exclusively as of July 29—that is, during the last five days of the term when a spate is almost guaranteed.

The seven canals of group II feed the middle reaches of the wadi (60% of the irrigated area) and receive water August 3–September 13. Usually, the main spates occur during this time span. The water is apportioned by upstream priority, with the exception of canals nos. 6 and 7, which have to share the water

43 For a comparison with the schedule of distribution of the base flow after implementation of the Tihāma Development Project, see Bonnenfant 1997, 19.

equally. Their intakes are the lowest within this group, and the areas irrigated by them regularly suffer from lack of water.

The four canals of group III have the right to allocate water to the low-lying field systems September 14–October 18, generally speaking a time of decreasing probability of spates in the wadi. Therefore, this is the land that is most frequently left fallow. If there is a weak spate, 50% of the *sayl* is distributed to canal no. 1, while nos. 2 and 3 receive equal shares of the rest. If the spate is strong, upstream priority is applied. From October 14 onwards, a family in the lower part of the area has the exclusive right to the water. During this period the spates often do not reach their land at all. This is why they regularly sell their title to farmers farther upstream.[44]

At the beginning of the 1970s, a total of 16,400 ha were irrigated in this way in the upper, middle, and lower reaches of the Wādī Zabīd.[45]

In addition, the aforementioned fields created through land reclamation in the low-lying areas along both sides of the wadi bed (*jalal*, sg. *jalla*, see section 1) have the right to use water from the *sayl* whenever it is available. Normally, this is the case during weak spates since the typical barrage that diverts the water from the wadi bed onto the field is of small size (*manṣūb*, see section 1) and is washed away immediately by a strong spate. In 1971, 3% of the land in the Wādī Zabīd (530 ha) was irrigated in this way.

The names of the individual canals (*shuruj*) in the Wādī Zabīd under the traditional system before implementation of the Tihāma Development Project are listed in table 4 (left column). The locations of the areas irrigated by the canals on the north and south banks of the Wādī Zabīd are shown on a map originally produced by ʿAbd al-Jawād al-Ṣulayḥī as part of his 1981 MA thesis (fig. 36).[46] Al-Ṣulayḥī distinguishes not only between the upper, middle, and lower reaches but also between the north—and the south-bank field systems, which makes the map quite detailed. See table 4 (second column from left) for a list of the names that are found on the map. The names are identical to those

44 Apparently, the family established their privileges through successful arbitration in disputes over water; see Tesco-Viziterv-Vituki 1971, 12; and Bonnenfant 1997, 19. With regard to what has been said earlier about water-allocation rights being tied to the land, the family's practice of selling their water rights independently of the land poses an interesting legal question; see Haddash 2001, 332–338.

45 In comparison, at the time of its greatest extent, the ancient oasis of Mārib covered an area of 9,600 ha; see Case Study 1, section 1.1.

46 Al-Ṣulayḥī, ʿAbd al-Jawād ʿAbd al-Ṣamad Thābit. 1981. "Jughrāfiyā al-maḥāṣīl al-zirāʿiyya al-raʾīsiyya fī l-Yaman." MA thesis, Ain Shams University, Cairo (unpublished). A reproduction of the map can be found in Shaqliyya 1992, 422. I am indebted to Jābir ʿAbduh ʿAlī ʿAbdallāh, Department of History, University of al-Ḥudayda, Yemen, for having drawn my attention to the map and for providing information on the author.

TABLE 4 Names of main canals and systems of contiguous fields irrigated by them (*shuruj*) in the Wādī Zabīd

Before implementation of Tihāma Development Project (as per Tesco-Viziterv-Vituki)	Before implementation of Tihāma Development Project (as per al-Ṣulayḥī)	According to tax register of fourth Rasulid sultan, al-Mu'ayyad Dāwūd (r. 1296–1321)	Operating during reign of seventh Rasulid sultan, al-Ashraf Ismāʿīl (r. 1377–1400)
group I (upper reaches): 1. al-Bunay 2. al-Barrī 3. al-Jirba 4. al-Manṣūrī 5. al-Rayyān 6. al-Baqar	group Ia (upper reaches, north bank): 1. al-Bunay 2. al-Barrī 3. al-Manṣūrī 4. al-Rayyān 5. al-Baqar group Ib (upper reaches, south bank): 1. al-Rawḍa 2. al-Jirba	north bank: 1. al-Bunay 2. Silsilī or Bulbulī (?) 3. Mubarraj 4. al-Māwī 5. Nābiṭ[a] 6. al-Rayyān 7. Majraʿa 8. Maʿẓab[b] 9. al-Jurayb 10. Mandab ʿAbbās 11. al-Minqāz (?) 12. Abī l-Rūm[c]	upper reaches (without claim to completeness):[f] 1. al-Mā'wī 2. al-Baqar 3. al-Rayyān 4. Nābiṭ[a] 5. Mubarraj 6. al-Naqḍ 7. al-Badānī[d] other canals mentioned (location not specified): 1. al-Minqāz[g] 2. Abīra[h]
group II (middle reaches): 1. al-Māwī 2. al-Yūsufī 3. al-ʿIbrī 4. al-Jarhazī 5. al-Jurayb 6. al-Bīra 7. al-Nāṣirī	group IIa (middle reaches, north bank): 1. al-Māwī 2. al-Yūsufī 3. al-Jurayb 4. al-Nāṣirī group IIb (middle reaches, south bank): 1. al-Ibrī [*sic*] 2. al-Jarhāzī [*sic*] 3. al-Bīra	south bank: 1. al-Naqḍ 2. al-Badānī[d] 3. al-Jarhazī 4. Abīra[e] 5. al-Dabra only mentioned on list of irrigation areas (not on map): 1. al-Baqar 2. al-Adīb	

TABLE 4 Names of main canals and systems of contiguous fields irrigated (cont.)

Before implementation of Tihāma Development Project (as per Tesco-Viziterv-Vituki)	Before implementation of Tihāma Development Project (as per al-Ṣulayḥī)	According to tax register of fourth Rasulid sultan, al-Muʾayyad Dāwūd (r. 1296–1321)	Operating during reign of seventh Rasulid sultan, al-Ashraf Ismāʿīl (r. 1377–1400)
group III (lower reaches): 1. al-Sharʿabī 2. al-Maḥraqī 3. al-Ḥaram 4. Wādī al-ʿAyn	group IIIa (lower reaches, north bank): 1. al-Sharʿabī 2. ʿAyn group IIIb (lower reaches, south bank): 1. al-Maḥrūqī [*sic*] 2. al-Ḥaram		

SOURCES: NAMES BEFORE THE IMPLEMENTATION OF THE TIHĀMA DEVELOPMENT PROJECT ARE DRAWN FROM TESCO-VIZITERV-VITUKI 1971, 6–11, AND FROM AL-ṢULAYḤĪ'S 1981 MAP OF THE CANAL SYSTEM IN THE WĀDĪ ZABĪD, REPRODUCED IN SHAQLIYYA 1992, 422. NAMES IN THE TAX REGISTER ARE DRAWN FROM JĀZIM 2008, 18 AND 387 (MAP), AND 41–44 (LIST OF IRRIGATION AREAS); NAMES DURING THE REIGN OF AL-ASHRAF ISMĀʿĪL ARE DRAWN FROM AL-KHAZRAJĪ/REDHOUSE, TRANS., AND ʿASAL, ED. 1906–1918.

a According to the local informants, today's name is al-Manṣūrī. See also Jāzim 2008, 43, n. 3.
b "Maʿḍab" on list of irrigation areas (Jāzim 2008, 42).
c "Maʾqam Abī l-Rūm" on map (Jāzim 2008, 387); it is a barrage built across the Wādī Zabīd, at its lower end, to raise the water to the level of the surrounding fields.
d According to the local informants, today's name is al-ʿIbrī. See also Jāzim 2008, 42, n. 2.
e According to the local informants, the names Bīra and Abīra were, and still are, used interchangeably.
f Al-Khazrajī/Redhouse, trans., and ʿAsal, ed. 1906–1918, 2:153 (Eng. trans.), 5:174 (Ar. text).
g Ibid., 2:243 (Eng. trans.), 5:272 (Ar. text "al-Minqār").
h Ibid., 2:253 (Eng. trans.), 5:284 (Ar. text). See also note e.

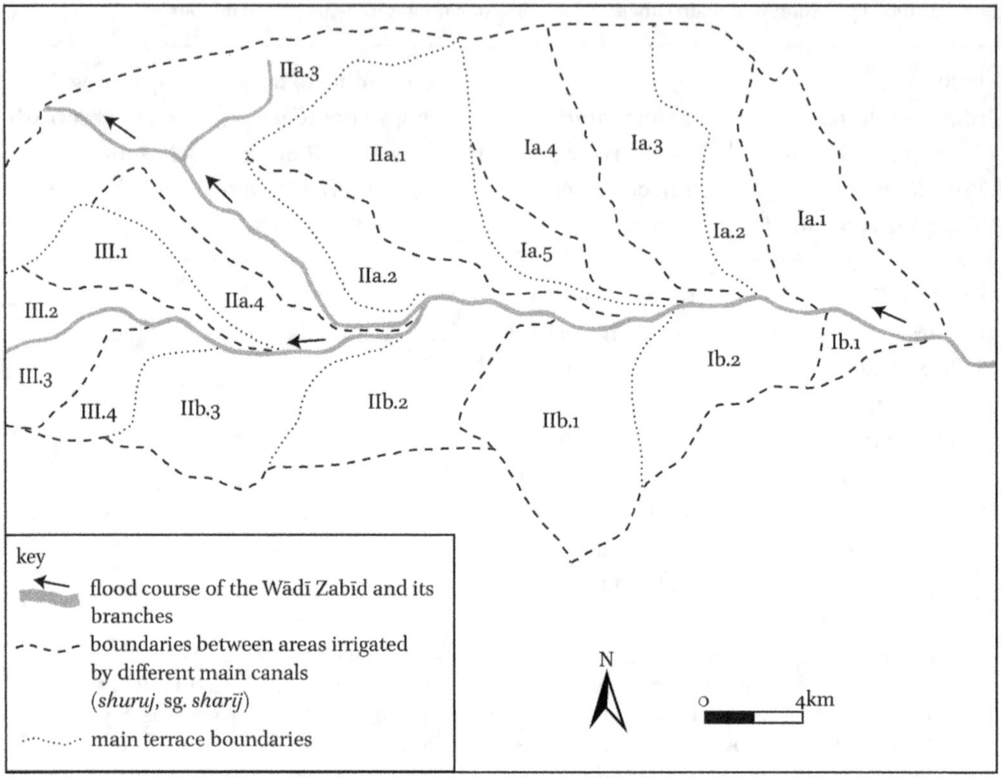

FIGURE 36 Map of the canal system in the Wādī Zabīd as documented before implementation of the Tihāma Development Project (adapted from reproduction of al-Ṣulayhī's map in Shaqliyya 1992, 422)

given by Tesco-Viziterv-Vituki, with the exception of one additional small irrigation area, al-Rawḍa, which belongs to group I on the southern bank.

Traditionally in the Wādī Zabīd, the *shaykh al-sharīj* (supervisor of an agricultural area irrigated by a specific *sharīj*)[47] bears responsibility for enforcing the law. He also ensures that repairs and maintenance work are carried out and collects the required monetary charges from the farmers, where applicable. The farmers pay according to the size of their irrigated land and its position within the canal system. Each canal is divided into an upper, middle, and lower section, and the ratio of payment is 4:2:1. The water-allocation supervisor also settles disputes among the farmers of the *sharīj*, or with those of other areas. His own experience as a farmer, together with his personality and ability to provide leadership, qualifies him for the job. The *shaykh al-sharīj* is recognized in

47 Tesco-Viziterv-Vituki 1971, 14; Serjeant 1988, 148; and Bonnenfant 1997, 15.

shāfiʿī law as a significant appointment, and its origin in *ʿurf* is acknowledged. Field-to-field distribution of the water within a *sharīj* is managed by the farmers themselves, not by the *shaykh al-sharīj*, and this tends to favour the large landowners who regularly take advantage of their stronger social standing over the smallholders in terms of the actual duration of irrigation.[48]

3.3 *Origins of the Water-Allocation Rights in the Wādī Zabīd*

To address the question of the origin of the unique water-allocation rights in the Wādī Zabīd, we may turn to some interesting legal disputes concerning water. According to local tradition, the earliest reference to the water rights in Zabīd is connected to the name of Shaykh Ismāʿīl al-Jabartī, who is given credit for establishing the law before 1473, the year of his death.[49] Serjeant has raised the possibility that al-Jabartī was only arbitrating, using the law rather than formulating it.[50]

In his 1995 MA thesis on "*Sharīʿa* and customary law in Zabīd," Nadim Salameh presents new evidence on the subject: a previously undocumented manuscript.[51] This text was drawn up around 1700 to settle, once again, conflicts arising from irrigation practices in the Wādī Zabīd. The author is the *shāfiʿī* scholar Muḥammad ibn Ziyād al-Waḍḍāḥī (d. 1722/3).[52] Clearly, quarrels among the farmers over maintenance and repair work prompted him to write

48 Tesco-Viziterv-Vituki 1971, 5 and 14–15; see Maktari 1971, 29–30 and 61–63, for the situation in the Laḥj Governorate.

49 Tesco-Viziterv-Vituki 1971, 5, citing Ibn al-Daybaʿ in general but without exact reference.

50 Serjeant 1974, 28 and n. 17. Unfortunately, the identity of Ismāʿīl al-Jabartī remains contentious. Tesco-Viziterv-Vituki (1971, 5) refers to the year 1473 as the year of his death as given by Ibn al-Daybaʿ, and the latter informs us that the man in question was Ismāʿīl ibn Muḥammad ibn Ibrāhīm al-Jabartī; see Ibn al-Daybaʿ/Chelhod, ed. 1983, 142–143. However, Ibn al-Daybaʿ does not provide any evidence for al-Jabartī's participation in shaping a new water law in the Wādī Zabīd. Serjeant (1974, 28) refers to an Ismāʿīl al-Jabartī who died in 1404. He cites as the source for the year of al-Jabartī's death his biography in al-Sharjī's *Ṭabaqāt al-khawaṣṣ* (1903). But Serjeant also notes that the latter text does not mention al-Jabartī's involvement with the water law in Zabīd either. Preference may be given to Tesco-Viziterv-Vituki's reference, in particular since the report was prepared by Abdulkarim al-Eryani (personal communication, May 2005), co-director of the Tihāma Development Project at the time and an expert in his field as well as in the history of Yemen. Keall and Hehmeyer (1998, 34) make a possibly erroneous connection with an Ismāʿīl ibn Abī Bakr al-Jabartī who died in 1470, as reported in Ibn al-Daybaʿ/Chelhod, ed. 1983, 139.

51 It was originally copied in Zabīd by Renate Cedzich during her field research in the second half of the 1980s and was bequeathed after her death to the University of Mainz. Salameh was granted permission to use the manuscript as the basis of his MA thesis. See Salameh 1995, "Vorwort."

52 Ibid., "Nachtrag September 1997" and 13.

the treatise.[53] Interestingly, he refers back to the unique Wādī Zabīd water law as originally having been codified in response to violent disputes among the farmers of the Wādī Zabīd during the term of office of Qāḍī Muwaffaq al-Dīn ʿAlī ibn Abī Bakr al-Nāshirī. He was a judge (*qāḍī*) in Zabīd from 1391 to at least 1400—that is, under the reign of the seventh Rasulid sultan, al-Ashraf Ismāʿīl I (r. 1377–1400).[54] It is at this time that the transition from *sharīʿa* to *ʿurf* was made.

In his text, al-Waḍḍāḥī makes a second reference to an earlier work that he used as a source. It was written by the *shāfiʿī* scholar Kamāl al-Dīn Mūsā ibn Aḥmad al-Dijāʿī and completed in 1570 under the title "The clear statements about the infamous actions which happened in the Wādī Zabīd."[55] Al-Dijāʿī describes twenty instances that he considers unlawful since they circumvent the customary rules in force. The infamous actions reveal interesting details with regard to farmers' practices and attempts to manipulate the system, mostly in order to increase the volume of irrigation water. For instance, case 1 reports that barrages were extended and made more solid (i.e., more permanent), thereby diverting more water than a particular canal (*sharīj*) was entitled to. In case 2, the small, normally temporary, dikes of mud and brushwood that were built across an irrigation canal to direct the water into a field unit were constructed with bricks and mortar to ensure that no water would pass to the next unit of fields in succession.[56] The tone of al-Dijāʿī's text is clear: the existing customary water law is valid and binding for everyone. Farmers are reminded to adhere to it.

It is obvious, then, that the new water law did not solve all the issues. But although this water law was challenged several times, it was consistently upheld, and it has, in fact, been in use in the Wādī Zabīd with minor modifications until

---

53 Ibid., 58–67.

54 Ibid., 41–42 and "Nachtrag September 1997." According to al-Khazrajī/Redhouse, trans., and ʿAsal, ed. 1906–1918, 2:194 and 286 (Eng. trans.), 5:220 and 316 (Ar. text), al-Nāshirī replaced his brother as *qāḍī* in Zabīd in 1391 by order of the seventh Rasulid sultan, al-Ashraf Ismāʿīl I. He held this position until at least 1400, the year of the sultan's death, which marks the end of al-Khazrajī's history.

55 For this work, see Salameh 1999, 137–142. It contains also a number of references to irrigation terminology, as does al-Waḍḍāḥī's text.

56 Interestingly, the term that is used in case 2 is *jisr*, not *maʿqam*, while in case 1, the term referring to the barrage is *maʿqam*. It is, of course, a well-known phenomenon that irrigation terminology is often ambiguous and tends to vary. Or may one consider the possibility that the learned legal scholar was not perfectly familiar with the irrigation terminology?

the present.[57] Clearly, the reason for its overall success is related to the fact that customary law is shaped by the farmers themselves—in fact, many generations of them. It is the farmers who are intimately familiar with the physical reality, most importantly the variability of the *sayl*'s timing and volume, rather than a legal authority who is removed from the harsh day-to-day conditions of irrigated agriculture in an arid environment. Combining upstream priority based on codified Islamic law (*sharīʿa*) with calendar-date allocation according to customary law (*ʿurf*) was seen as a way of spreading the risks as well as the bounty more equitably across the system.

At the expense of efficiency, the new water-allocation rights established relative stability in the Wādī Zabīd. This is also reflected in a remarkable continuity regarding property names from medieval to premodern and even modern times. Table 4, second column from the right, lists the *sharīj* names given in the tax register of the fourth Rasulid sultan, al-Muʾayyad Dāwūd (r. 1296–1321). See the right column for the names that al-Khazrajī refers to in various contexts in his *al-ʿUqūd al-luʾluʾiyya* as operating during the reign of the seventh Rasulid sultan, al-Ashraf Ismāʿīl I (r. 1377–1400); unfortunately, al-Khazrajī does not produce a comprehensive list. Interestingly, seven of the nineteen *shuruj* cited in the Rasulid tax register operated under the same name until the implementation of the Tihāma Development Project in the late 1970s. These are Sharīj al-Bunay, al-Māwī, al-Rayyān, al-Jurayb,[58] al-Jarhazī, Abīra/Bīra,[59] and al-Baqar. Most probably, all nine canals mentioned by al-Khazrajī for the time of the seventh Rasulid sultan are included in the tax register of his great-grandfather. This kind of continuity reflects the fact that irrigation schemes require major investment. New names appeared occasionally through the sponsorship of later rulers, as in the example of Sharīj al-Manṣūrī, the former Sharīj Nābiṭ that was renamed after the ninth Rasulid sultan, al-Manṣūr ʿAbdallāh (r. 1424–1427). Sharīj al-Badānī is another example of a main canal operating during Rasulid times that was renamed (Sharīj al-ʿIbrī).[60] Thus, nine of the nineteen main canals in the Rasulid tax register can be identified in the Wādī Zabīd before the implementation of the Tihāma Development Project.

57 Tesco-Viziterv-Vituki 1971, 15. The report recommends that the implementation of the impending Tihāma Development Project should leave the existing water-allocation rights for spate irrigation in the Wādī Zabīd unchanged (vii and 16), and this is what happened, more or less.

58 Serjeant (1988, 146) claims that Sharīj al-Jurayb was known from the early twelfth century onwards, unfortunately without citing a reference.

59 According to the local informants, these two names were, and still are, used interchangeably.

60 According to the local informants; see also Jāzim 2008, 42, n. 2, and 43, n. 3.

3.4 *Conclusion*

Documents dealing with customary law have a unique value as historical sources. Unlike the chronologies and official histories that tend to focus on the ruling elite and often do little more than list noteworthy events from political incidents to natural disasters, these legal agreements usually follow first a dispute and then its settlement. Often written in colloquial Arabic, they include explanations of and comments on what happened and why, thus giving remarkable insights into the practical problems of daily life and the solutions applied to the individual situations.[61] Since the rules relating to sharing water ultimately depend on the available water resources, customary water law varies from one region to another. Therefore, many *ʿurf*-based texts concerning water were primarily of local interest. Few copies were made, and even fewer have survived.[62] For Zabīd, they reflect community-based water management—a practical, bottom-up approach as opposed to a theoretical, top-down approach. Even though as written documents they were produced by people who were literate, these texts are nevertheless as close to the farmers and their actual problems as one can hope to find.

61 See Serjeant 1979, 99–100.
62 See Salameh 2001, 215.

CASE STUDY 3

The Importance of the Agricultural Hinterland to Pre-Ottoman Zabīd

In 1997 the Canadian Archaeological Mission began a building-conservation program in the Iskandariyya mosque in Zabīd (fig. 37). The unexpected exposure of an inscriptional panel flanking both sides of the prayer niche (*miḥrāb*) opened a new window into the medieval landscape of the Wādī Zabīd and the relationship between the city and its agricultural hinterland. Much to the surprise of everybody involved, the inscription was found to contain not the expected Koranic text but rather a dedication of a long forgotten charitable endowment (*waqf*, pl. *awqāf*) in support of a religious college (*madrasa*, pl. *madāris*), an institution usually associated with a mosque. The *waqf* lists individual properties of farmland in the wadi and measures of harvested grain derived from them.[1] The text reflects the principles of water management described in Case Study 2, while at the same time underlining the significance of Zabīd's role as a centre of learning.

1 The Text

The right panel, measuring approximately 0.80 × 1.90 m, consists of a single slab of stone bearing the first thirteen lines of the inscription. The text continues with lines 14–27 on the similarly sized left panel, which is made up of five separate pieces of stone (fig. 38). The stone slabs are perhaps repurposed backs of tombstones. Instead of the text being cut into the stone, the background was carved out around the letters and then painted in gilt; hence the protruding text stands out clearly against it.

The text is reproduced in the appendix to this chapter. It starts with the *basmala*, the formula "In the name of God, the Merciful, the Compassionate" that constitutes the first line of all the chapters of the Koran except the ninth. Invocation of the *basmala* should precede any important act in a Muslim's life, calling for its divine blessing and consecrating it.[2] The execution of the *basmala* (line 1, see fig. 39) accentuates what, according to tradition, Caliph ʿUmar

1 For some preliminary observations, see Keall and Hehmeyer 1998, 33–35 and 47.
2 Carra de Vaux 1960, 1084a.

 DOI:10.1163/9789004387713_007

FIGURE 37
Al-Iskandariyya mosque-*madrasa* in Zabīd

FIGURE 38 The two panels of the *waqf* inscription on either side of the prayer niche in the prayer hall of the Iskandariyya mosque-*madrasa*

FIGURE 39 Close-up of right panel, lines 1–3

(r. 634–644) said to his scribe: "Lengthen the *bāʾ*, make the teeth of the *sīn* prominent and round off the *mīm*."[3]

After the *basmala*, the reading gets much more difficult, so difficult indeed that the local dignitaries and religious authorities initially suggested that the text was not written in Arabic but in Ottoman Turkish. Clearly, the stone carver faced a considerable challenge fitting the entire text onto the two panels. With the exception of the first line carrying the *basmala*, what looks like one line of text consists, in fact, of three lines of writing, and frequently the reading starts at the bottom, continues at the top of the line, and then jumps to the middle. A rather liberal approach is taken to diacritical signs, which appear where there is a space, and in most instances they are shared by the adjacent letters and words. In order to facilitate the reading, a pencil rubbing was made, which forces one to follow the cut marks of the carving systematically and thus allows for easier identification of the individual letters. Nevertheless, reading the text and making sense of it required a combined effort involving local and external specialists.[4]

3 Ibid., 1084a–b.

4 I am particularly grateful to ʿAbd al-Karīm Aḥmad al-Sālimī, director of the Zabīd Office of the General Organization of Antiquities and Museums (GOAM), and Abdallah Ghouchani, Organization for Iranian Cultural Heritage, Tehran, Iran.

1.1 *Iskandar, Sponsor of the* Waqf

The text goes through several lines of self-claimed honorifics of the sponsor of the inscription, whose name is eventually given in line 8 as Iskandar min Barsbāy.[5] Since the date of the inscription is specified at its end (line 27) as the beginning of the month of Muḥarram 940/July 1533, we can take this man to be Iskandar ibn Sūlī, also known by his popular name as Iskandar Mawz, the last Lawandi officer to govern Zabīd, having seized power in 1530. He acquired a local reputation for justice and munificence and successfully ruled a coalition of Mamluks and Lawandis for six and a half years, much longer than any of his predecessors. He died in 1536 of natural causes.[6]

At first sight, the particle *min* in Iskandar's name is unusual and was, in fact, originally read as *bin* (son of), a mistake that can easily happen because of the similarity of the two words; *bin* is the normal manner of referring to a family relationship. However, the particle *min* is a familiar way in which the late Mamluks expressed an affiliation, with a whole range of possible relations to the *min* person.[7] Iskandar, for instance, seems to have appealed for legitimacy through using the name of Barsbāy, the first Mamluk ruler of Zabīd following the joint Mamluk-Lawandi invasion of 1516, who—like Iskandar—is described in the historical sources as enjoying popularity among his fellow Mamluks and Lawandis.[8]

Al-Nahrawālī, the sixteenth-century chronicler of Yemen, refers to Iskandar Mawz as having "built" (*banā*) a *madrasa* in Zabīd by the name of al-Iskandariyya,[9] and since the mosque has long been known locally by this name, it has generally been understood by scholars that the mosque was

5 A number of the honorific titles are adjectives ending in the suffix *-ī* (*nisab*, sg. *nisba*). While constructions such as *al-kabīrī* may seem grammatically unusual, the impressive list of Iskandar's titles copies, in fact, parts of formulas that were well known in the Middle East during the Mamluk period (1250–1517); e.g., Wiet (1932) 1984, 101–102 (cat. 3982); and Mols 2006, 304–311 (cat. 50, in particular 50/1 and 50/15). A variety of Mamluk objects bearing similar formulaic inscriptions have been sold by auction houses during the past decades, e.g., a candlestick auctioned by Sotheby's in 2007 (http://www.sothebys.com/en/auctions/ecatalogue/2007/arts-of-the-islamic-world-including-fine-carpets-and-textiles-l07222/lot.138.html, accessed December 2013). See also Björkman 1928, 112.

6 Al-Nahrawālī/al-Jāsir, ed. 1967, 56; and Blackburn 1971, 64. Blackburn vocalizes Iskandar's name as Mūz. The Canadian Archaeological Mission vocalizes it as Mawz; see Keall 1984, 54.

7 Ayalon 1975, 223–228 and 231–232. The author points out that use of the particle *min* for this purpose is exclusive to the Circassian Mamluk period (1382–1517) and can be observed increasingly frequently towards its end. It is important to note that the *min* person was in no way restricted to being a Mamluk's manumitting patron.

8 See Blackburn 1971, 41–43 and 45.

9 Al-Nahrawālī/al-Jāsir, ed. 1967, 58.

constructed by Iskandar.[10] On this basis, it was even dismissed rather cavalierly as being of no great interest in terms of Yemen's art and architecture because of its relatively recent date as an "Ottoman" mosque.[11] This misjudgment results from "not distinguishing between the somewhat unstable period, when the Lawandi mercenaries rivaled for power …, and the formally administered Ottoman occupation (after 1539)."[12]

As early as 1983 and 1984, Edward J. Keall suggested that the mosque was originally a fourteenth-century Rasulid construction.[13] His argument is based on an assessment of the building's multidomed architecture, its layout, and the style of its original painted ceiling decorations. The striking minaret is a later addition. This is clear from the join between mosque and minaret, where the two are attached but do not actually bond. The base of the minaret bears a dedicatory inscription, in which part of the name of Iskandar can be made out, together with a scarcely legible but nevertheless discernible date giving the month of Ramadan of the year 940/1533 that confirms the identity of the sponsor.[14] Keall speculates that the associated Iskandariyya attribution was likely due to a donation of monies made by Iskandar to the mosque, along with a renovation scheme and the building of the minaret.[15]

This suggestion was corroborated by the 1997 building-conservation program that allowed for careful scrutiny of the mosque structure and its interior, including excavation of a small sondage in the prayer hall. The results indicate that al-Nahrawālī's statement about Iskandar's having built the *madrasa* should not be taken literally. Instead, the association with Iskandar, the patron of the *madrasa*, should be connected with the later period of renovation and redecoration, when the existing mosque was turned into a religious college and lost its original attribution. This did not involve any changes to the architecture of the mosque itself, because accommodation was provided in a separate building. It simply meant that, in addition to being a place of prayer, the building became a place of teaching and studying.[16] But there is unequivocal evidence of changes made to the decorative scheme in the survival of part

10 See, for instance, Finster 1993–1996, 24 and n. 40. Finster errs twice by attributing the construction of the mosque to Iskandar and by stating that this happened around the year 1520.

11 Finster 1992, 135.

12 Keall 2001b, 221, n. 4.

13 Keall 1983, 59; and Keall 1984, 54.

14 For the minaret inscription, see Keall 1989a, 68. The reading of the date was slightly refined in light of the *waqf* dedication next to the prayer niche of the mosque.

15 Keall 1983, 59; and Keall 1984, 54.

16 See Finster 1993–1996, 25.

FIGURE 40 Decorative panel in the northeastern corner of the courtyard arcade of the Iskandariyya mosque-*madrasa*. The central part of the original Rasulid plasterwork has been replaced with a new inscription

of a panel lodged in the northeastern corner of the courtyard arcade (fig. 40). The style of design for this carved plasterwork is distinctly different from that found elsewhere in the building. Significantly, the central part of the panel was hacked out and replaced with a new dedicatory inscription. In other words, this was a retroactive modification of the earlier panel.

We find evidence of the sponsor in the more recent plaster style running along the wall on the south side of the courtyard. Figure 41 shows the letters *kāf–nūn–dāl* of Iskandar's name.[17] The same *kāf–nūn–dāl* can be observed on the base of the minaret. Lastly, the renovation included the insertion of the two stone panels with the *waqf* inscription on either side of the prayer niche. They were clearly fitted into the existing north wall of the mosque.

1.2 *The Agricultural Details*

The name Iskandar min Barsbāy as given in the first part of line 8 is followed by the specifics of Iskandar's sponsorship. He endowed as a *waqf*, in support of

17 Keall (2001b, 221–222) gives a brief description of what was involved in Iskandar's major renovation of the building.

FIGURE 41 The letters *kāf–nūn–dāl* of Iskandar's name, inscribed on the southern courtyard wall

the *madrasa*, the product of harvests derived from distinct tracts of land in the Wādī Zabīd. The inscription lists three areas named after the main canal irrigating each one: Sharīj al-Ṭāhirī, Sharīj Marḍī, and Sharīj al-Jurayb (see table 5). For each of these, the text identifies a number of individual properties, the sizes of which are given in *maʿād*, pl. *maʿāwud; maʿād* is the only term used, and smaller units are simply designated as "one-half," "one-quarter," or "one-eighth" of a *maʿād* (see fig. 42). For instance, the area of Zahab Suwayd covers "three *maʿād* and one-half *maʿād* and one-quarter *maʿād*" (lines 13–14).[18] The *maʿād* is still in use today in the Wādī Zabīd as a land measure that equals 0.35 ha.[19]

The portions of the harvest levied from these properties are listed in units of *mudd*, pl. *amdād, thumn*, pl. *athmān*,[20] and *zabadī*, pl. *azbūd* (see fig. 43).

18 In Arabic, proper fractions (numbers between zero and one) are expressed by means of unit fractions (fractions with unit numerators). E.g., "three-quarters" is written as "one-half and one-quarter." The only exception to this rule is "two-thirds," a fraction people felt comfortable with. This is a phenomenon that we observe as early as ancient Egypt; see Boyer 1989, 15–16. See table 6 below for examples from medieval Zabīd.

19 Tesco-Viziterv-Vituki 1971, 18–19. See also Grohmann 1933, 98.

20 The literal meaning of *thumn* is "eighth." Donaldson (1996, 34) explains why derivations from the root *th–m–n*, based on the number eight, are very common in the traditional

TABLE 5 The agricultural details of the *waqfiyya* of the Iskandariyya *madrasa* (1533)

Name of irrigation canal	Name of property	Levy from property in measures of capacity	Area of property (*maʿād*)
Sharīj al-Ṭāhirī	Ḍāḥiyat al-Dabwān	16 *mudd* + 2 *thumn* + 7½ *zabadī*	60 + ½
	Zahab al-Fārisī	8 *mudd* + 31 *thumn*	29
	Zahab Suwayd	2 *mudd*	3 + ½ + ¼
	Ḍāḥiyat al-ʿĀṣirī	3 *mudd* + 1 *thumn*	10 + ½
Sharīj Marḍī	Zahab Laʿbār	2 *mudd* + 17 *thumn*	13 + ¼
	al-Ṣanʿāniyya	12 *mudd* + 16 *thumn*	43 + ½
	al-Ṣafwānī	10 *mudd*	30 + ½
(?)	Ḍāḥiyat Ṭabʿāya	7 *mudd* + 21 *thumn*	25 + ½ + ⅛
Sharīj al-Jurayb	Ḍāḥiyat Hadhma	8 *mudd*	21
Total (inscription)		Σ=71 *mudd* + 2 *thumn*	Σ=237
Total (calculated)		Σ=68 *mudd* + 88 *thumn*+ 7½ *zabadī* =70¾ *mudd* + 7½ *zabadī*	Σ=237 ⅝

FIGURE 42 Close-up of left panel of the *waqf* inscription, lines 16–17, with *rubʿ maʿād* ("one-quarter of a *maʿād*") legible at the left end of line 16

metrological system of Yemen. This phenomenon has its origins in ancient South Arabia; see Sima 2000, 200, (1) and n. 96; and Maraqten 2014, 112.

All three terms denote measures of capacity for grain. However, specifying their respective sizes is a problem as measures varied widely from one location to another, and also over time.[21] For instance, al-Khazrajī, the fourteenth-century court historian of the Rasulids (d. 1410), states that in Ḥays, a town some 35 km south of Zabīd, levies were measured in "the *mudd* of the people of Yemen" ("mudd ahl al-Yaman"),[22] while according to the historian and religious scholar Ibn al-Dayba' (1461–1537), in Zabīd "the *mudd* of Zabīd" ("mudd zabīdī") was used.[23] In the city of Ta'izz, approximately 100 km southeast of Zabīd, "the *zabadī* of Ta'izz" ("zabadī ta'izzī") was stipulated by the local authorities.[24] The enormous variation in terms of size from one location to another during Rasulid times is best understood from a late-thirteenth-century royal ledger attributed to the second Rasulid sultan, al-Muẓaffar Yūsuf (r. 1250–1295). It contains detailed administrative and fiscal information,[25] including a long list of the sizes of the customary dry measures for a great number of cities and regions.[26] Surprisingly, Zabīd is not among them.

However, we find in al-Khazrajī's work a full description of the changing size of the *zabadī* in Zabīd from Ayyubid to Rasulid times. The *zabadī sunqurī* was named after Sayf al-Dīn Sunqur (d. 1212/13),[27] a high-ranking official under the Ayyubids, who set it at 240 dirhams, with the dirham being a unit of weight of goods in this context, not a unit of currency.[28] At an unspecified time, a later ruler raised it by a third to 320 dirhams, and this remained its size until towards the end of the reign of the fifth Rasulid sultan, al-Mujāhid 'Alī (r. 1321–1363),

21 The standard 1970 reference text by Hinz, *Islamische Maße und Gewichte umgerechnet ins metrische System* (Islamic weights and measures converted to the metric system), is not helpful for medieval Yemen; the same is true of Schultz 2003, 59–75. Donaldson (1996, 34) points out that the traditional measures of capacity, stemming from before the introduction of the metric system in the 1970s, "vary considerably throughout the Yemen and even within a small area" and gives ample examples throughout the article. See also Grohmann 1933, 97–100. Varisco (2015, 103) states that the size of the *zabadī* "differed according to both region and product." A confusing diversity of weights and measures already existed in ancient South Arabia, as described by Maraqten (2012, 107–116).

22 Al-Khazrajī/Redhouse, trans., and 'Asal, ed. 1906–1918, 1:231–232 (Eng. trans.), 4:276 (Ar. text).

23 Ibn al-Dayba'/Chelhod, ed. 1983, 192.

24 Al-Khazrajī/Redhouse, trans., and 'Asal, ed. 1906–1918, 4:148 (Ar. text). The text passage is not included in Redhouse's English translation. See also Varisco 2006, 171–172.

25 The manuscript was edited in two volumes by Jāzim (2003–2005). See Vallet 2010, 70–72, for general information on the text.

26 Jāzim 2003–2005, 1:340–349.

27 For information on Sayf al-Dīn Sunqur, see Smith 1978, 2:97 and 98.

28 See Hinz 1970, 1–4, for a general distinction between the dirham as a coin weight and a weight of goods.

FIGURE 43 Close-up of left panel of inscription, lines 18–19, with *sabʿat amdād* ("seven *amdād*") legible at the left end of line 19

when in 1359/60 the governor of Zabīd together with the local excise officer increased the *zabadī* by another 40 dirhams. Later excise officers increased it again and again, until the issue seems to have run completely out of control during the reign of the sixth Rasulid sultan, al-Afḍal al-ʿAbbās (r. 1363–1377). His successor, al-Ashraf Ismāʿīl I (r. 1377–1400), finally took charge of the problem and ordered in 1384 that the *zabadī* should be lowered to 400 dirhams. But a grain scarcity and rising prices later that year forced him to increase the *zabadī* to 500 dirhams. Al-Khazrajī ends the section with the remark that, at least during his lifetime, this size of the *zabadī* became permanent, much to the distress of the farmers.[29] Thus the weight of the *zabadī*—and the volume, because the weight of a body increases in proportion to its volume[30]—had more than doubled during some 150 years of Rasulid rule, with enormous fluctuations along the way. For the individual farmer, this meant a considerable increase in taxes on grain (taxes on staple crops were paid in kind) since a corresponding re-assessment quite clearly did not happen.[31]

---

29 Al-Khazrajī/Redhouse, trans., and ʿAsal, ed. 1906–1918, 2:159–160 (Eng. trans.), 5:180–181 (Ar. text).

30 Nontechnical language uses "weight" instead of "mass," which is the correct term with regard to the underlying principles of physics.

31 See Vallet 2010, 337–338, for an interesting interpretation of the passage with regard to inflation and favouring the interests of the urban population over those of the farmers.

At Iskandar's time, no doubt, the capacity of the *zabadī* as well as the *mudd* and the *thumn* had changed once again, but their actual sizes are not known. Fortunately, this does not apply to the ratio of the three measures of capacity to one another because the inscription lists not only the exact levies from each individual property and its respective area (lines 9–22) but also the sum totals of both the levies and the farmland (line 22). The sum total of the farmland is specified as 237 *maʿād*, while adding up the individual areas of the properties results in 237⅝ *maʿād*. What this means is that the figures given in the inscription are reliable, albeit with minor percentage differences.

For the harvest levies, the total amount is recorded as 71 *mudd* and 2 *thumn*. Adding up the individual levies results in 68 *mudd* and 88 *thumn* and 7½ *zabadī*. Comparison of these two sums allows us to deduce the following ratios of the three dry measures to one another:

- The *mudd* is the largest unit; *thumn* and *zabadī* are subordinate.
- The individual levies are listed in quantities of up to 31 *thumn*, which suggests that a *mudd* exceeds this figure. For the late Ayyubid period, we have the account of Ibn al-Mujāwir (d. before 1250), probably a businessman, who visited Zabīd and gives a detailed report on the different grain measures that were in use at the time:[32]

  1 *mudd* = 32 *thumn*[33]
  1 *thumn* = 32 *zabadī*[34]
  1 *zabadī* = 1 *mann*
  1 *mann* = 2 *raṭl*
  1 *raṭl* = 120 dirhams
  1 dirham = 13 *qīrāṭ*

  If we add up the individual levies under the assumption that the ratio of 1 *thumn* = 1⁄32 *mudd* was still valid during Iskandar's time, we obtain a calculated total of 70¾ *mudd* and 7½ *zabadī*.
- The size of the *zabadī*, then, is similar to the size of the *thumn*.

Unfortunately, the two lines following the sum total of the levies and the farmland are illegible (lines 23–24). It seems that they were deliberately hacked out so that the content that caused this act of destruction would never be known. The text ends with an invocation to God to accept the endowment and words

32 Ibn al-Mujāwir/Smith, trans. 2008, 114. Varisco (1994, 164–165), also referring to Ibn al-Mujāwir, erroneously gives the ratio 32 *thumn* = 32 *zabadī*; see Ibn al-Mujāwir/Löfgren, ed. 1951–1954, 1:89.

33 This is also the ratio that is given in the list of dry measures from the royal ledger attributed to al-Muẓaffar Yūsuf, dated to the end of the thirteenth century: 1 *mudd sunqurī* = 32 *thumn*. See Jāzim 2003–2005, 1:343.

34 The royal ledger attributed to al-Muẓaffar Yūsuf gives the following ratio: 1 *thumn* = 10 *zabadī sunqurī*. See Jāzim 2003–2005, 1:343.

of warning to anybody who may violate its conditions (lines 25–27), followed by the date, which is squeezed into the left corner of line 27. Towards the end of the text, the stone carver had clearly run out of space.

1.3 *A Comment on Style*

The text contains a number of peculiarities and errors with regard to orthography and grammar: the sound plural masculine (nominative) sometimes ending in *-īn* in place of the classical *-ūn*, the dual (nominative) ending in *-ayn* in place of *-ān*, faulty gender polarity of the numbers three to ten with the counted noun, and faulty case of the counted noun following the numbers eleven to ninety-nine, among others. The linguistic features are characteristic of what is called "Middle Arabic," "that literary form of the Arabic language which is a mixture of non-classical Arabic elements and pure classical Arabic elements."[35] Interestingly, the errors are not consistent throughout the text. This seems to imply that the person who drafted it made a deliberate effort to apply the rules of classical Arabic, the language of the Koran, but lacked their perfect command, and hence he confused them with the simplified rules of the colloquial language. The deficiencies in classical Arabic are somewhat surprising since the inscription was made for a religious institution.

However, they would not have diminished the overall impact of the inscription. The insertion of the stone panels on either side of the prayer niche—that is, in the most prominent position—guaranteed that Iskandar's *waqf* dedication was seen by everybody who came to the *madrasa* to pray or to study. The gilt paint added to the fact that it could not be overlooked, and the illegibility of the writing would have lent it even greater authority. While the bold statement made in the inscription can be seen in propagandistic terms, one also gets the sense that Iskandar was genuinely concerned about the upkeep of the *madrasa* bearing his name.

2 Practical Implications of the *Waqf*

The inscription lists canals and the tracts of land irrigated by them using the terminology that is familiar from Case Study 2. One of the canals, Sharīj al-Jurayb, was known in Rasulid times (see table 4). It is a canal that belongs to the middle reaches of the Wādī Zabīd. As is implied by its name, Sharīj al-Ṭāhirī was established by the successors of the Rasulids, the Tahirids (1454–1517). In today's Wādī Zabīd, it is no longer considered a main canal (*sharīj*) but rather

35 Smith 1996, 328.

a secondary canal (*sāqiya*) that irrigates farmland in the lower reaches of the wadi, below Sharīj al-Maḥraqī on the south bank. Sharīj Marḍī is today also a secondary canal in the lower reaches and supplies farmland on the north bank with water. Thus, with the exception of a rather small property in Sharīj al-Jurayb (Ḍāḥiyat Hadhma, measuring 21 *maʿād*), the land that Iskandar bequeathed to the *madrasa* was not exactly of high quality and would not have produced satisfactory harvests in years when the *sayl* was weak (see Case Study 2, section 3.2). Interestingly, al-Khazrajī mentions that when the fifth Rasulid sultan, al-Mujāhid ʿAlī (r. 1321–1363), set up a *waqf* for a mosque that he had constructed in Mecca, he deliberately chose land for it from the upper, middle, and lower reaches of the Wādī Zabīd in order to minimize the risk of a crop failure.[36] It was the land in the upper reaches that guaranteed a harvest even when the *sayl* did not materialize because this region had access to the base flow that would only dry up during a prolonged drought (see tables 2 and 3).

Iskandar donated the land with the levies derived from it specified in the inscription as an endowment (*waqf*) to the *madrasa*. Strictly speaking, there are different types of *waqf*, with the relevant one in this context being the *waqf khayrī* (public-good *waqf*), a "charitable endowment for the support of public municipal and religious institutions."[37] At its core are the principles of charity (one of the five pillars of Islam) and perpetuity. The dedication of property that guarantees provision of significant services is permanent; *waqf* property can neither be transferred nor alienated, and only the income generated by it is assigned to charity. The conditions of the *waqf* are supposed to be immutable. While the different branches and schools of Islam vary with regard to certain details, most of them agree that a valid *waqf* has three essential components: first, a founder of mature and sound mind; second, his or her specific property that is to be converted to a *waqf*; and third, one or more beneficiaries (persons or public utilities, such as mosques or schools). The actual founding of a *waqf* requires a declaration of intent by the founder that is usually written down in a document, the *waqf* deed (*waqfiyya*). It needs to be legally authenticated, and while it is usually kept with a judge, in some cases the text is found carved on walls of buildings, as in the Iskandariyya mosque-*madrasa*.[38] A *waqfiyya* is an important legal document, especially at times when the parties concerned have to be reminded of the conditions of the endowment.

---

36 Al-Khazrajī/Redhouse, trans., and ʿAsal, ed. 1906–1918, 2:55–56 (Eng. trans.), 5:68 (Ar. text).

37 Shatzmiller 2001, 46 and n. 7.

38 The institution of the *waqf* and its legal framework are discussed in Sait and Lim 2006, 147–153; and in Peters 2002, 59a–63b.

Throughout *al-ʿUqūd al-luʾluʾiyya*, al-Khazrajī regularly refers to the establishment of endowments for religious institutions, many of them in Zabīd. The sponsors were the rulers themselves, their relatives (including female relations), high-ranking court officials, or wealthy individuals. Usually, al-Khazrajī makes a general statement that the *waqf* was sufficient to cover the maintenance expenses and thus to guarantee perpetuation of the institution.[39] Ibn al-Daybaʿ also writes about endowments, most importantly in the context of his chief patron, the fourth (and last) Tahirid sultan, al-Ẓāfir ʿĀmir II (r. 1489–1517). In 1494, al-Ẓāfir ʿĀmir II bequeathed land as a *waqf* to the Grand Mosque (*jāmiʿ*) in Zabīd, with the levies amounting to 100 *mudd* per year.[40] Unfortunately, Ibn al-Daybaʿ states neither the location nor the area of the land.

Detailed information on the *awqāf* of Zabīd is given in the tax register of the fourth Rasulid sultan, al-Muʾayyad Dāwūd (r. 1296–1321), compiled at the end of the thirteenth century. Nineteen endowments in the Wādī Zabīd are itemized, each one with the name of the mosque or *madrasa* that they supported and the respective levies of grain in measures of capacity (see table 6).[41] The list reflects the full range from very wealthy to not so wealthy. As in the example of the *waqfiyya* of the Iskandariyya *madrasa*, the grain would have consisted of different varieties of millet and white and red sorghum, the food staples in Zabīd both in the past[42] and present. The sum total of the endowments is some 607 *mudd* in levies, and the tax register lists these under the rubric "exempt from taxation" because *waqf* property is not subject to land tax. By comparison, the sum total of the actual tax revenues collected by the state in kind—that is, in grain—in the Wādī Zabīd is specified as "7,238 *mudd* and 14 *thumn* and one-half and one-third and one-eighth."[43]

Almost a century later, in 1393, the seventh Rasulid sultan, al-Ashraf Ismāʿīl I (r. 1377–1400), ordered a survey of Zabīd's religious buildings. It noted some 230 mosques and *madāris*, a figure that reflects the city's significance as a centre of learning.[44] Zabīd acquired a reputation for teaching and scholarship

39 E.g., al-Khazrajī/Redhouse, trans., and ʿAsal, ed. 1906–1918, 1:232, and 2:101 (Eng. trans.), 4:276–277, and 5:118–119 (Ar. text).

40 Ibn al-Daybaʿ/Chelhod, ed. 1983, 191–192. The dry measure is specified as *mudd zabīdī*.

41 Jāzim 2008, 35–37.

42 Ibn al-Mujāwir/Smith, trans. 2008, 112; and Varisco 1994, 165.

43 Jāzim 2008, 31. Four endowments that consist of date-palm groves are itemized separately. Levies from them were paid in cash (24–30).

44 Al-Khazrajī/Redhouse, trans., and ʿAsal, ed. 1906–1918, 2:216 (Eng. trans.), 5:244 (Ar. text). The *madrasa* as an institution of higher education was introduced to Yemen by the Ayyubids (1173–1228); see Finster 1998, 25. However, not a single *madrasa* from this time has survived in Yemen. The earliest extant examples were built by the Rasulids.

TABLE 6 List of the *awqāf* in the Wādī Zabīd at the end of the thirteenth century

Name of mosque (M) or *madrasa* (*m*)	Levy in measures of capacity
al-Dār al-Saʿīd al-Shamsī (M)	14 *mudd* + 20 *thumn* + ½ + ⅓
al-Dār al-Saʿīd al-Wāthiqī (M)	70 *mudd* + 14 *thumn* + ½
al-Ashāʿir (M)	23 *mudd* + 31 [*thumn*][a] + ⅔ + ¼
Jihat al-Ṭawāshī Iqbāl al-Juzarī (*m*)	92 *mudd* + 26 *thumn* + ½ + ⅓
Sayf al-Dīn Atābik (M)	8 *mudd* + 26 *thumn* + ⅔ + ¼
al-Ṭawāshī Tāj al-Dīn Badr (*m*)	255 *mudd* + 28 *thumn* + ½ + ⅓ + ½ × ⅛[b]
al-Jamālī (?)[c]	30 *mudd* + 1 *thumn* + 2 *qīrāṭ* + ½
al-Amīr Badr al-Dīn Muḥammad ibn ʿAlī al-Hukārī (*m*)	25 *mudd* + 8 *thumn* + ⅓
Khayl Khān (M)	9 *mudd* + 25 *thumn* + ⅔
al-Ḥājja Khayzurān (M)	7 *mudd* + 10 *thumn* + ¼ + ⅙
al-Faqīh Sirāj al-Dīn Abū Bakr ibn Daʿʿās (*m*)	31 *mudd* + 26 *thumn* + ½
al-Ḥājja Umm ʿAfīf (*m*)	20 *mudd* + 1 *thumn* + ⅛ + 1 *qīrāṭ*
al-Muqrī al-Jalawī (M)	4 *mudd* + 30 *thumn* + ½ + ⅓ + ⅛
Ḥāfat al-Widn (M)	1 *mudd* + ⅔ + ⅛
al-Minjāra (M)	4 *mudd* + 22 *thumn* + ½ + ¼ + ⅛
al-Qāḍī ʿAmrān (M)	1 *mudd* + 29 *thumn* + ⅛ + 2 *qīrāṭ*
al-Qāḍī Abū Bakr ibn ʿAbdallāh (M)	1 *mudd* + 25 *thumn* + ½ + ¼
Umm ʿAlī ibn Aḥmad (M)	16 *thumn* + ⅙
al-Ḥalabī (M)	28 *thumn* + ⅔ + ⅛
Total (calculated)[d]	Σ=607.14 *mudd* + 5½ *qīrāṭ*

SOURCE: TAX REGISTER OF THE FOURTH RASULID SULTAN, AL-MUʾAYYAD DĀWŪD (R. 1296–1321) (SEE JĀZIM 2008, 35–37)

a No measure of capacity specified, with the logical one being the *thumn*.

b The Arabic language has no specific terms—derived from the name of the denominator—for fractions smaller than one-tenth. Wherever possible, in these cases the denominator is broken down into aliquot parts lower than ten, so that a fraction smaller than one-tenth is expressed by a fraction (or fractions) of a fraction. For instance, one-sixteenth is expressed as "one-half of one-eighth" (Arnaldez and Massignon 1963, 406; and Sabra 1971, 1140b–1141a).

c No specification made.

d With the ratio of 1 *mudd* = 32 *thumn*.

not only in the religious sciences but also in the mathematical sciences.[45] Al-Khazrajī refers to a local scholar in the second half of the fourteenth century who was skilful in arithmetic and astronomy, and to the appointment of a professor of inheritance law,[46] a discipline that included—besides the legal provisions—the mathematics of inheritance. After all, determination of the individual shares according to Islamic law involves complex mathematical procedures. The appointment was made at the mosque where al-Khazrajī himself held the position of professor of Koran recitation. Ibn al-Daybaʿ informs us about the subjects that he studied in Zabīd as a young student in the second half of the fifteenth century. Besides Arabic and jurisprudence, he specifies the following mathematical disciplines: arithmetic, algebra, surveying, and inheritance law.[47]

The individuals affiliated with a *madrasa* typically included one or more professors, students, an imam (leader in the congregational prayer), a muezzin, a precentor, a caretaker, a teacher, and orphans learning the Koran.[48] Of course, it would be interesting to know how many individuals were supported by Iskandar's *waqf*. According to Varisco, who refers to an unpublished manuscript from the thirteenth century housed today in a private library in Ṣanʿāʾ, the *zabadī* was the amount of grain consumed by an individual in the period of one month.[49] Unfortunately, since the location to which this statement refers is not known, it has little value in the context of this chapter and does not help to estimate the number of people who could have been fed from Iskandar's endowment. A unique *waqfiyya* from Rasulid times specifies that during the reign of the sixth sultan, al-Afḍal al-ʿAbbās (r. 1363–1377), the imam of the Salāma *madrasa* in Taʿizz received a stipend of 50 *zabadī* of grain per month; the muezzin, 30.[50] With regard to the *waqfiyya* of the Iskandariyya *madrasa* in Zabīd, it has been described above that the capacity of the measures changed frequently, and we simply have no information on their sizes during Iskandar's time.

45 For an interesting comment on the alleged invention of algebra in Zabīd, see King 2012, 223–231.

46 Al-Khazrajī/Redhouse, trans., and ʿAsal, ed. 1906–1918, 2:154 (not translated properly into English) and 178–179 (Eng. trans.), 5:175 and 202 (Ar. text).

47 Ibn al-Daybaʿ/Chelhod, ed. 1983, 218.

48 For some rather generously sponsored *madāris* in fourteenth-century Zabīd, see al-Khazrajī/Redhouse, trans., and ʿAsal, ed. 1906–1918, 2:101 (Eng. trans.), 5:118–119 (Ar. text).

49 Varisco 1994, 164.

50 Sadek 1990, 150–154. For further information on various monthly stipends and fees measured in *zabadī* in the Taʿizz region towards the end of the thirteenth century, see Varisco 2015, 102–103. Varisco refers to details given in the aforementioned royal ledger attributed to the second Rasulid sultan, al-Muẓaffar Yūsuf.

3 Conclusion

Building conservation may lead to surprise discoveries, as in the case of Iskandar's forgotten *waqf* in support of a *madrasa*. In the course of our work on the Iskandariyya mosque, the text spelling out the details of the endowment emerged from underneath many layers of whitewash that had obscured it from view. It gives us a rare glimpse of the semi-autonomous Mamluk-Lawandi military regime in Zabīd between 1517 and 1539, for which there is otherwise little concrete evidence. From the inscription, we gain detailed information ranging from the measures of capacity and surface area that were in use at the time to linguistic challenges and the applications of arithmetical rules for expressing proper fractions. The *waqfiyya* also proved to be an unexpected reward insofar as it meshes nicely with what we have learned about *sayl* irrigation in the Wādī Zabīd from studying the archaeological remains, the historical sources, the terminology, and the traditional practices. It fits the overall picture of continuity in the Wādī Zabīd. Finally, the inscription underlines the importance of successful management of the water resources in the wadi for sustenance of Zabīd's economic, religious, and scholarly life.

4 Appendix: The *Waqfiyya* of the Iskandariyya *Madrasa* by Iskandar min Barsbāy, Dated 1533

The following text is as found in the inscription (including peculiarities and errors in orthography and grammar). Letters in square brackets ([x]) indicate additions; obvious errors are marked by "[*sic*]"; "[…]" is lost text that cannot be reconstructed; "[?]" draws the reader's attention to an obscure word.

١ بسم الله الرحمن الرحيم الحمد لله رب العالمين
٢ وصلى الله على سيدنا محمد وآله وصحبه وسلم هذا ما وقف
٣ وحبس وتصدق وسبل على هذه المدرسة المباركة هو
٤ سيدنا ومولانا المقر الكريم العالي المولوي الكبيري العالمي
٥ العادلي المالكي ا[*sic*]الملكي المجاهدي المرابطي العولي [؟]
٦ العالي صاحب الاا[*sic*]قاليم الشامية والبلدان اليمنية ملك البر والبحر
٧ ذو الفضل والكرم والجود والنعم والسيف والقلم
٨ والبيد [و]العلم اسكندر من برسباي اوقف لمصالح هذه المدرسة
٩ الاراضي الا[*sic*]تي ذكرها بالصورة المشروحة وهم ضاحية
١٠ الدبوان بشريج الطاهري الغلة ستة عشر مدا وثمنان وسبعة ازبود
١١ [و]نصف زبدي المعاود ستون معاد ونصف زهب الفارسي بالشريج المذكور
١٢ الغلة ثمانية امداد وواحد [و]ثلثون ثمنا المعاود تسعة وعشرون معاد
١٣ زهب سويد بالشيج المذكور الغلة مدين المعاود ثلاثة
١٤ معاود ونصف معاد وربع معاد ضاحية العاصري بالشريج المذكور
١٥ الغلة ثلاثة امداد وثمن المعاود عشر معاود ونصف زهب لعبار بشريج
١٦ مرضي الغلة مدين وسبعة عشر ثمنا المعاود ثلاثة عشر معاد وربع معاد
١٧ الصنعانية بالشريج المذكور الغلة اثنا عشر مدا وستة عشر ثمنا المعاود
١٨ ثلاثة واربعون معاد ونصف معاد الصفواني بالشريج المذكور الغلة
١٩ عشرة امداد المعاو[د] ثلاثون معاد ونصف معاد ضاحية طبعاية الغلة سبعة امداد
٢٠ وواحد [و]عشرون ثمنا المعاود خمسة وعشرون معاد ونصف معاد وثمن معاد
٢١ ضاحية هذمة بشريج الجريب الغلة ثمانية امداد المعاود واحد [و]عشرون
٢٢ معاد جملة المعاود مائتين وسبعة [و]ثلاثين معاد الغلة واحد [و]سبعين مدا [و]ثمنين
٢٣ […]
٢٤ […]
٢٥ وقفا صحيحا شرعيا خالصا لوجهه الكريم تقبل الله ذلك وثمنه [و] كرمه فمن بدله
٢٦ بعدما سمعه فانما اثمه على الذين يبدلونه فلعبة الله ولعبة الملائكة وا[ل]لاعنين
٢٧ والناس اجمعين عليه وكتب بتاريخ مستهل شهر محرم الحرام سنة اربعين وتسعمائة

CASE STUDY 4

Engineered Water Systems in the Wādī Zabīd

The infrastructure on which *sayl*-irrigated agriculture in the Wādī Zabīd depended in the past was mainly built with unmanufactured, locally available, and non-durable materials. For instance, traditional diversion barrages consisted of heaped-up earth, coarse gravel, and boulders, all taken from the flood channel; they were often reinforced with remnants from agriculture and natural vegetation, such as palm fronds and brushwood. Spillways that accommodated drops in elevation were an exception insofar as they were constructed with baked brick laid in lime mortar. However, not all buildings related to water in the Wādī Zabīd were devices for diverting and distributing the seasonal *sayl*. Over the course of our fieldwork, a number of structures were identified that turned out to belong to water systems not directly dependent upon the wadi spate.[1] They were built to be durable, and after falling out of use, they were buried under the rising sediments. There are some scattered references to such engineered water systems in the historical sources, but there were no obvious sites for archaeologists to look for them. It was the farmers of the Wādī Zabīd who showed us the remains that they had fortuitously discovered in their fields. Serendipity therefore played a major role by providing us with the starting points for our archaeological work.

1 Plaster-Lined Underground Canal (ZHB 30)

While bulldozing the sediments some 2 km east of Zabīd to level a new field, a farmer exposed two parallel stretches of brick masonry. Excavation revealed a plaster-lined underground canal (ZHB 30)[2] some 35 cm wide on the inside, running east–west (see fig. 44 for location). It became clear from the section that it had been constructed in a trench dug in open ground. Footings consisting of one or two courses of brick were laid with lime mortar to the full width

1 Preliminary observations on these systems were published in Keall 1993, 475–477; Hehmeyer and Keall 1993, 25–27; and Hehmeyer 1995, 45–54.

2 The Canadian Archaeological Mission's grid-system code is explained in Keall 1991, 85, n. 4. In order to record the excavations, three-letter site designations, e.g., ZHB, were used. Each of them represents a 1 km^2 project grid that includes 100 grid units measuring 100 × 100 m each. They are numbered from 0 to 99. The underground canal described here is located in grid unit 30.

 | DOI:10.1163/9789004387713_008

of the trench. Portions of the original canal walls of up to nine courses of brick were found preserved, the inner faces waterproofed with lime plaster (locally known as *qaḍāḍ*;[3] see fig. 45). A distinct protrusion in the north wall at about 90 cm from the bottom surface inside the canal has been interpreted as the remnant of the springing of an arch that originally capped it. In spite of the fact that the inside is extremely narrow, several layers of increasingly coarse lime plaster, which were recorded on the inner faces, support the notion that the canal was replastered from time to time, either by a small man or a boy who would have been able to work inside it.

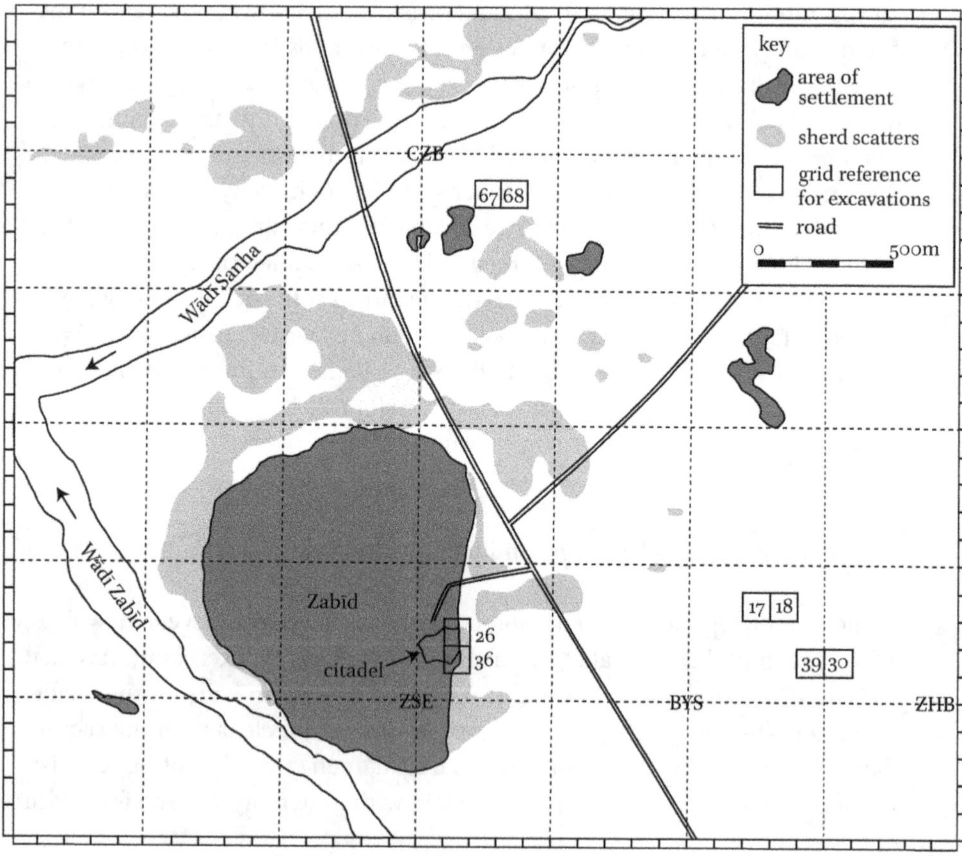

FIGURE 44 Excavation sites of pre-Ottoman engineered water systems in and around Zabīd, with three-letter site designations and grid-unit numbers

3 The composition, preparation, and application of *qaḍāḍ* are explained in Part IV, Case Study 7, sections 2 and 3.

FIGURE 45
Underground canal (ZHB 30; 8 cm scale)

At some point higher canal walls were built, with the new south wall set back up to 15 cm from the original one, resulting in a wider canal. The lower part of the initial construction was left intact and must have continued to function properly because the new plaster lining was only taken to about 40 cm above the bottom. The weakness of the new structure, however, appears to have occurred at the join between the old and the new part because it was replastered many times. The new walls are preserved up to 145 cm in height from the original bottom, unfortunately without an indication of how high they might have reached. The bricks used in the rebuilt canal are of a smaller, thinner, and harder (i.e., high-fired) type than those in the original construction; because of their characteristic properties, they have been termed "small-tablet" type (see table 7). Building the canal higher and wider might have become necessary to increase the volume of water in general or to accommodate the rising level of the sediments that would have progressively restricted the volume of water flowing out at the far end under the force of gravity.

Significantly, set deliberately in the footings of the canal was a palm-sized pottery sherd displaying fresh breaks (fig. 46).[4] The breaks indicate that it was not a small fragment that had bounced around for years, becoming abraded and its edges rounded. It was set in the mortar of the footings to fill a gap in the bricks, a practice that has been observed numerous times in the work of the builders using traditional techniques in Zabīd. The pottery sherd would have been handed to the builder by his assistant, picked up from their immediate environs. If there had been any delay, the builder could have easily chipped a piece of appropriate size from a brick. In all likelihood, this sherd was from a pot broken during the course of the work on site. In terms of the Zabīd pottery typology, the sherd represents a transitional phase beginning in the tenth century—that is, in the second half of the Ziyadid period—and later replaced by a new tradition that started in the eleventh century after the end of Ziyadid rule (see table 1). Hence, the pottery sherd gives an approximate date for the construction of the original canal.[5] It allows us also to assign the same date to the large, fat, soft (i.e., low-fired) type of brick that was used to build the original canal and that is distinctly different from the much smaller, thinner, and harder small-tablet type of the reconstruction phase.

Zabīd's famous historian, Ibn al-Dayba', describes a similar canal of his day: "There is an abundantly flowing spring, the water comes from the east in an underground canal until it approaches the city, then it emerges at ground surface."[6] Concerning the original construction of the canal, Ibn al-Dayba' continues: "The first one who tapped the aforementioned spring and built the canal for it and led it to the city was the Qāḍī al-Rashīd … Aḥmad … Ibn al-Zubayr …, the author and poet from al-Aswān. He was unique in his age in the science of engineering and in the legal sciences."[7]

Ibn al-Zubayr, the Egyptian water engineer who spent some of his life in Yemen, eventually returned to his native country, where he died in 1167/8.[8]

---

4 A comparable piece is illustrated in Ciuk and Keall 1996, plate 95/30-b. While this work provides the basis for the Zabīd pottery typology, certain details such as the assigned dates have quite naturally been refined since its publication in 1996; see Keall 2012, 131–132.

5 The slightly revised date is based on personal communication with Edward J. Keall, August 2011.

6 Ibn al-Dayba'/Chelhod, ed. 1983, 48.

7 Ibid.; the statement is repeated on page 73.

8 Ibid., 48. According to the thirteenth-century biographer Ibn Khallikān, the year of Ibn al-Zubayr's death is 1166. See Ibn Khallikān/de Slane, trans. 1842,1:143.

TABLE 7 Zabīd brick typology (part 1)

Grid-system code	Archaeological feature	Brick size (cm)	Date	Basis for date	Remarks
ZHB 30	plaster-lined underground canal, original construction	20 × 20 × 5.5–6	10th–11th c.	transitional-phase sherd set in brick masonry	bricks soft, low-fired
ZHB 30	plaster-lined underground canal, rebuilt	13 × 15 × 3.5 (small-tablet type)	first c. of Rasulid rule up to mid-14th c.	BYS 18, pipe beds (phases 1 and 2)	bricks hard, high-fired
BYS 39	stilling and distribution drum	—	—	—	—
BYS 18	pipe beds (phases 1 and 2)	14 × 14 × 3.5 to 15 × 15 × 4.5 (small-tablet type)	first c. of Rasulid rule up to mid-14th c.	globeware sherds set in brick masonry	bricks hard, high-fired
BYS 18	open canal (phase 3)	24 × 25 × 5 to 26 × 26 × 5.5; and 14 × 14 × 3.5 to 15 × 15 × 4.5 (small-tablet type)	—	—	small-tablet type of bricks are recycled and used both whole and as fragments
BYS 18	brick kiln: grill ribs	20.5 × 21 × 4–4.5 (and two smaller bricks: 16.5 × 17 × 3.5)	—	—	—
BYS 18	brick kiln: wall set down into the fire chamber	13 × 13 × 3.5 to 13.5 × 13.5 × 4 (small-tablet type)	first c. of Rasulid rule up to mid-14th c.	BYS 18, pipe beds (phases 1 and 2)	bricks vitrified; probably later modification of fire chamber
BYS 18	brick kiln: bricky zone inside kiln, overlying mud surface and cobblestone base	18 × 18.5 × 4 to 18.5 × 19 × 4.5 (and one smaller brick: 14 × 15 × 4)	—	—	bricks underfired, soft
CZB 67–68	pipe bed	14 × 14.5 × 3.5 to 15 × 15 × 4 (small-tablet type)	first c. of Rasulid rule up to mid-14th c.	BYS 18, pipe beds (phases 1 and 2)	—

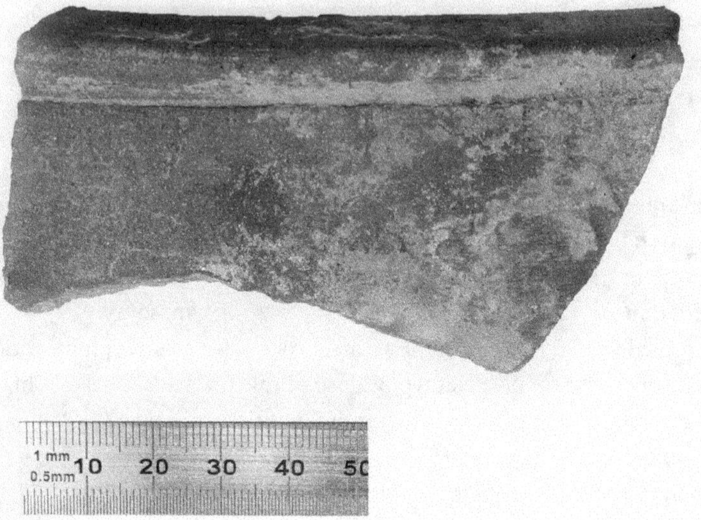

FIGURE 46 Transitional-ware sherd from the footings of the underground canal (ZHB 30)

Other writers relate even earlier evidence for engineered water-delivery systems; for example, the tenth-century geographer al-Muqaddasī reports that "Ibn Ziyād had a channel led to the town."[9] The reference is to the beginnings of Zabīd and the person who founded it in 820.

There may certainly be remains of other canals in the environs of Zabīd besides the one excavated by us. It is also not clear as to where this particular one was directed, nor does its dating precisely match the textual references. The archaeological evidence documented by us is nevertheless significant as, even without direct correlation, it lends credence to the statements made in the texts.

What was the source of the water feeding the system? First of all, a canal of this kind could not possibly cope with the muddy spate; it would have become clogged over time. In addition, the investment was enormous and only justifiable by providing a year-round flow of water. Conveniently, Ibn al-Dayba' mentions a spring east of Zabīd, so we may be confident that the water source was located towards the foothills, some 20 km to the east of the city, where it would have been possible to tap into the groundwater or into the perennial base flow of the wadi. In either case, it would have been particle-free, clear water.

9 Al-Muqaddasī/Collins, trans. 1994, 82. See the Introduction to this book, section 5, on "canal" vs. "channel."

2 Stilling and Distribution Drum (BYS 39)

A short distance to the west of ZHB 30, excavation exposed a round brick structure with an internal diameter of 1.65 m and its masonry preserved up to a height of 1.05 m (BYS 39, see fig. 47). It is lined on the inside with lime plaster, as is a stepped canal (with the lowest step and the edge of the second step preserved) that enters the structure some 45 cm above its bottom. At the same height in the opposite wall, three glazed earthenware pipes terminate side by side (see table 8 for the datum heights). The pipes were laid in a bed of baked-brick masonry extending from the round structure. A slight recess in the wall around the opening of the pipes became necessary to accommodate the fact that their ends are not flush with the circular wall. The setback was neatly plastered after construction was completed. Upon excavation, the left and the middle pipe were found deliberately plugged with lime plaster. The third pipe on the right was open, retaining traces of a sleeve of lead sheeting with which the end had been extended out from the recess. A solitary fourth pipe with a considerably larger diameter was positioned to the left and some 20 cm higher than the group of three. It was clearly a later addition, the opening for it hacked through the brick wall and crudely finished. Unfortunately, because of the location of the archaeological remains right next to a high field bank and a farmyard, it was not possible to excavate the pipe bed.

FIGURE 47 Stilling and distribution drum (BYS 39; 50 cm scale)

TABLE 8 Datum heights (above sea level) of engineered water systems ZHB 30, BYS 39, BYS 18, and CZB 67–68

Grid-system code	Archaeological feature	Component	Datum height (m)[a]
ZHB 30	plaster-lined underground canal, original construction	bottom of canal, inside	107.23
BYS 39	stilling and distribution drum	bottom of drum, inside	107.85
BYS 39	stilling and distribution drum	cluster of three pipes centred	108.30
BYS 39	stilling and distribution drum	bottom of right pipe out of cluster of three pipes	108.26
BYS 39	stilling and distribution drum, later addition	fourth pipe centred	108.50
BYS 39	stilling and distribution drum, later addition	bottom of fourth pipe	108.44
BYS 39	stilling and distribution drum	bottom of lowest step of stepped canal	108.31
BYS 39	stilling and distribution drum	bottom of second step of stepped canal	108.51
BYS 18	pipe bed, phase 1	top pipe of cluster of three pipes centred (east profile of trench)	107.43
BYS 18	pipe bed, phase 1	bottom of top pipe of cluster of three pipes (east profile of trench)	107.38
BYS 18	pipe bed, phase 2	fourth pipe centred (east profile of trench)	107.64
BYS 18	pipe bed, phase 2	bottom of fourth pipe (east profile of trench)	107.59
CZB 67–68	pipe bed	top pipe centred	98.85
CZB 67–68	pipe bed	bottom of top pipe	98.82

a The Project datum point of 100 m was established on the foot of one of the electricity pylons on the east side of Zabīd that lies close to the 100 m above sea level contour line.

The function of the circular structure was to stall the water flowing in through the stepped canal before it would flow out through the pipes. The device operated as a stilling and distribution drum. As with the underground canal ZHB 30, the water source was most probably the perennial base flow of the wadi or groundwater; the muddy spate would have quickly blocked the pipes. This was confirmed when the lime packing in the left and the middle pipe was removed and no sediment deposit was found inside. While it might be tempting to speculate about a physical connection between canal ZHB 30 and drum BYS 39,[10] site surveying has established that the datum heights of the two structures negate this possibility. The canal bottom is more than 0.5 m below the bottom of the drum, and the water would simply not have flowed into the drum (table 8).

3 Pipe Beds, Open Canal, and Brick Kiln (BYS 18)

A tentative connection can, however, be made between the drum and the pipe system extending from it and other portions of pipe beds found during subsequent field seasons. The archaeological remains of BYS 18 are located some 175 m to the northwest of the stilling and distribution drum BYS 39 (see fig. 44) and consist of a three-phase engineered water system. In the first phase, a pipe bed of baked-brick masonry was constructed on open ground along a dry stream bed. It housed three glazed earthenware pipes.

Later, another pipe bed was built adjacent to, but slightly higher than, the earlier one, carrying a single and considerably larger pipe (fig. 48). A thin but distinct line of separation between the two construction phases is clearly visible in the brick masonry. By phase two, the ground had risen by 30 cm, and the more recent pipe bed was also built on what was then open ground. In both phases, the small-tablet type of bricks was used. Fragments of pottery (globular cups for drinking, decorated with thinly scratched lines) were found embedded in the mortar of the masonry. As with the transitional-phase sherd from the footings of the canal ZHB 30, this "globeware" pottery can be associated with a particular period of construction. It corresponds approximately to the first century of Rasulid rule up to the mid-fourteenth century.[11]

The pipelines consisted of individual cylindrical earthenware pipe sections, with one end tapered and fitted with a circular exterior flange (fig. 49). The tapered "male" end was designed to fit into the wider "female" end of its

10 See Keall 1993, 477.

11 For a drawing of a globeware cup, see Hehmeyer 1995, 50, fig. 3 (*bottom*).

FIGURE 48 Sequence of glazed earthenware pipes (three below, one above) and partial remnant of open canal (*top right*) (BYS 18; 8 cm scale)

FIGURE 49 Joint between two pipe sections with lime mortar (BYS 18)

neighbour. It allows one to determine the direction of flow. Each individual pipe section is hand formed, with distinct imprints of the potter's fingers on the inside, and shows traits of having been made from a slab of clay that was modelled into a cylindrical form, possibly around the potter's forearm. The flange around the male end was affixed separately. The sidewalls of the individual pipe sections are 1–2 cm thick. The units of phase 1 measure 41.5–45 cm in length (phase 2: 44–45 cm) and have an internal diameter of some 4.5 cm at the male end (phase 2: 8–9 cm) and 8.5 cm at the female end (phase 2: 11–12 cm). The pipe units were glazed on the interior and exterior with a transparent yellowish-to-turquoise glaze. Lime mortar was applied around the joint, which was then wrapped with a piece of linen; the impression of the woven fabric is well preserved in the mortar.[12] The pipelines were embedded in brick masonry. While the joint was not entirely waterproof, the seepage through the encapsulating lime mortar would have been minimal.

The pipes BYS 18 were directed not west to the city of Zabīd but to the northwest, as were the pipes BYS 39. It is entirely feasible that the two were connected, for the gradient between BYS 39 and BYS 18 amounts to 0.5% (ca. 90 cm in 175 m; see table 8).[13]

At some point a third structure was added to the water system. This phase is marked by a change in brick type and mode of operation. The archaeological remains suggest that an open canal was built alongside the existing pipe beds (fig. 48). It was built with two different brick types. The small-tablet-type bricks of the earlier two phases are found both whole and as fragments, indicating

---

12 The fabric is assumed to be linen based on identification of similar-looking material used to seal lengths of lead pipe in the excavations ZSE 36. See Case Study 5, section 6 and fig. 66.

13 In terms of the engineering standards at the time, the significance of this figure should not be overinterpreted. As Hodge (1992, 115) points out commenting on Roman water systems,

"Unlike the normal masonry channel, which was designed to run only half-full and under gravity flow, pipes ... were intended to run full. This in turn means that a pipeline would run under at least nominal pressure. We are not here speaking of the high pressures generated in a siphon or any of the other locations that have so worried modern commentators. All that is meant is that, unlike an open channel, a pipe need be laid only roughly level. Minor bumps and dips in the line are of no importance. The water will surmount them without trouble, so long as the general tendency of the line is gently downhill, and though this will inevitably cause some changes in pressure within the pipes, the changes will be so small and the pressure so low as to be insignificant and easily contained both by terracotta pipes and even their plastered joints. In theory, this means that a pipeline is probably easier to lay out than a conventional channel. For there is much greater tolerance in establishing the level."

See the Introduction to this book, section 5, on "canal" vs. "channel."

that they were recycled from another building and that during this process a number of them were broken. In addition, considerably bigger whole bricks, measuring up to 26 × 26 × 5.5 cm, were employed. They seem to represent the brick type that was manufactured locally at the time and used new in the construction.

The demise of the open canal of phase three must have occurred when a flash flood caused a washout underneath it, leaving behind a telltale deposit of sand and gravel. A large part of what was most probably the outer canal wall toppled over to the side. Only a few diagnostic sherds were found to date this context, but the absence of any recognizable Ottoman-period pottery is significant.[14] We may therefore safely deduce that the system had fallen into disuse by Ottoman times.

Attempts were made to trace the piped system BYS 18 farther through the fields, and during this work, a sounding resulted in the exposure of another fragmentary stretch of the two pipe beds as well as a substantial brick kiln (both also BYS 18). The kiln was originally quite large, measuring more than 4.80 m in length as excavated. It operated on the basis of an arched grill covering a fire chamber (fig. 50). The bricks of the grill ribs measure 20.5 × 21 × 4–4.5 cm (table 7). The kiln was built over a mud surface overlying cobblestones that mark its base level. Several repairs were observed, and surfaces had been vitrified and distorted by heat in the normal manner of kiln use. It appears as though a later modification was made that consists of a solid L-shaped wall of vitrified bricks, each course separated by a thin band of sandy dirt. Since the bricks are laid to a face and bonded in the corner angle, it does not seem likely that this was a stack of overfired bricks left behind in the kiln. Instead, it suggests that a wall of bricks was set down into the kiln as a somewhat crude alteration of the fire chamber. These bricks are of the distinct small-tablet type.

The earliest deposit at the bottom of the fire chamber included some whole but underfired, unusable bricks in a loose, crumbly matrix that belong once again to a different type (see table 7). From the brick types recorded in the kiln, we can infer different stages of its use, with the aforementioned modification of the fire chamber datable to the first century of Rasulid rule up to the mid-fourteenth century.

In the sounding, the fragmentary stretch of the two pipe beds was found to run over the top of the kiln, which had either fallen into disuse at the time of construction or had at least been put out of commission by the construction work for which it may have produced the bricks. The difference in height between the surface on which the lower pipe bed was built and the mud layer

14 The characteristic Ottoman pottery is described in section 6 below.

FIGURE 50 Brick kiln (BYS 18; 1 m scale)

over which the kiln was constructed amounts to 1.5 m. Unfortunately, although very hospitable in giving access to his field, the enterprising landowner started large-scale bulldozing of the terrain in preparation for a new irrigation plantation shortly after the initial excavations were finished. This made it impossible for us to determine whether the fire chamber of the kiln might have been dug at least partly into the ground. It would have been interesting to know how much of the 1.5 m difference in height corresponds to the time between the construction of the kiln and the installation of the pipe bed.

4 Pipe Bed (CZB 67–68)

Northeast of the city (see fig. 44 for location), another well-preserved stretch of pipe bed was fortuitously exposed by construction machines at two different sites 120 m apart, with a gradient of 0.64% (77 cm in 120 m). The pipe bed was built of whole bricks of the small-tablet type (table 7), and it encased three lines of glazed earthenware pipes set in the same triangular formation as observed in BYS 18. The masonry bed itself was constructed in cut-and-fill style; that is, a trench was dug in open ground to build the pipe bed that was covered with the spoil after completion. The cutline from construction is clearly visible in section. When broken open, the pipes were found not to contain any sediment deposit. This confirms once again that it was not muddy *sayl* water that fed the piped systems. Since the datum height of the pipes at CZB 67–68 is several metres below the heights of the pipe beds described earlier, there was no connection with these systems (see table 8).

5 Conclusion

Like the pipes BYS 18 and BYS 39, the pipes CZB 67–68 are directed not to Zabīd but north past the city. They seem to have served a suburban site. Archaeological remains confirm that garden estates existed, for instance, 1 km north of Zabīd, and also on the east side of the city.[15] The travel account of Ibn Baṭṭūṭa, who visited Zabīd around 1330, states that the city "lies amid luxuriant gardens with many streams and fruits, such as bananas and others."[16] As we have seen above (Part II, Introduction, section 1), the Rasulid rulers took great personal interest in agriculture and horticulture, including the cultivation of exotic crops and flowers. For this, the biannual *sayl* of limited duration was simply not a sufficient water source; however, the engineered water systems detailed here allowed for the required year-round irrigation. Even though the water would have been of potable quality, the people of Zabīd preferred drinking their own well water, as reported by Ibn al-Daybaʿ.[17]

Zabīd's folklore has given us the colourful story of the lad who, in order to win the hand of the princess, is tricked by the sultan into building a canal all the way from the mountains to the sea. Needless to say, the suitor is too old and unsightly by the time he has finished the project, which covers a distance of some 45 km. In another version, the young man himself decides to build the canal in order to ease his young wife's homesickness by having fresh flowers from the mountains floated every day to the hot and dry coastal plain. Such stories often have some factual basis, and they were probably created to explain what in hindsight appears to be an engineering project on an enormous scale with regard to the resources required for construction. Besides the pipe sections, huge quantities of bricks had to be produced.

Different types of bricks were recorded over the years in various archaeological contexts. They can provide revealing information about the structures in which they were used, in particular with regard to assigning them approximate dates. For instance, the small-tablet type, and with it the rebuilding phase of canal ZHB 30, can be dated to the first century of Rasulid rule

15 Keall 1989a, 67; and Hehmeyer 1995, 52.
16 Ibn Baṭṭūṭa/Gibb, trans. 1962, 2:366.
17 Ibn al-Daybaʿ/Chelhod, ed. 1983, 48.

up to the mid-fourteenth century; the large, fat, soft type from the footings was employed during the transitional phase from the tenth century (i.e., in the second half of the Ziyadid period) to the eleventh century after the end of Ziyadid rule. Both types are easily recognizable in other contexts due to their characteristic properties, especially the small-tablet bricks that were widely used in the engineered water systems in the Wādī Zabīd. The archaeological remains described here confirm that the first half of the Rasulid era was a time of economic prosperity, when substantial investment was made, but that the engineered water systems in the Wādī Zabīd originate from considerably earlier times. Clearly, they were well maintained, so that Ibn al-Daybaʿ could point out that at his time the underground canal originally built by Ibn al-Zubayr in the twelfth century irrigated "all the gardens outside the city and within it."[18] However, the complete absence of Ottoman pottery in the parts of the environs of Zabīd described here strongly suggests that the water systems feeding the land had fallen into disuse by the mid-sixteenth century. See the postscript below for the characteristic Ottoman pottery.

6 Postscript: Excavation of a Post-Ottoman Spillway

During the 2007 season of excavations, a stepped spillway (*jisr*) (fig. 51) was excavated in the upper reaches of the Wādī Zabīd, close to the village of Maḥall al-Shaykh in the area irrigated by Sharīj al-Baqar.[19] It accommodated a drop in elevation between the irrigated farmland and a discharge gully. The arched spillway was built through the bank of Sharīj al-Baqar with baked brick set in lime mortar. The bricks were laid in splays of increasing verticality, with a brick cut to a wedge shape inserted at the apex of the arch. The total height of the stepped spillway measures 2.20 m on the inside. In order to strengthen the arch, it was covered with baked-brick fragments laid in mud mortar. A 2 cm thick layer of lime plaster coated the outside.

A 12 cm insert on either side narrows the spillway before it reaches the first of three steps. The insert has three pairs of sockets (at heights of 44 cm, 87 cm, and 128 cm above the bottom) to hold three bars for sluice gates that allowed for control of the water flow. The water dropped down three steps (1.40 m in total). Traces of the springing of the arch are visible in the brickwork above the first step. The entire face of the inner walls of the unit is coated with lime

18 Ibid.

19 For Sharīj al-Baqar, see Case Study 2, table 4 and fig. 36 (I.a.5).

FIGURE 51
Stepped spillway excavated in the Wādī Zabīd, with three sockets visible (1 m scale)

plaster (*qaḍāḍ*). On the steps it has a pebble-rich matrix to withstand water erosion.

During excavation, the sluiceway was exposed on its exterior down to its footings, which start 3.30 m below the top of the arch. The rather irregular footings consist of baked brick in mud mortar. A later repair on the western side of the unit was carried out in brick laid in mud mortar and covered with pebble-rich lime plaster. The bricks have what is for Zabīd a rather unusual rectangular shape (ca. 22–24 × 11–12 × 6 cm) and are crudely made; some are partly melted from overfiring.

The pottery recovered from the lower layers of dirt just outside the unit contained sherds with a whitish body and a dark leaf-green glaze. The whitish clay is a characteristic of pottery made in Ḥays, a town some 35 km south of Zabīd.

Manufacture of the white-bodied ware started in the fifteenth century[20] and reached its peak with what we refer to as "fine Ḥays pottery," using transparent green (or brownish-yellow) glazes over an almost white body.[21] It is associated with the first Ottoman occupation of Yemen and started to be produced some time after the arrival of the Ottomans in 1538. Interestingly, the earliest products have the finest quality, which had already started to decline towards the end of the sixteenth century. Although the tradition continued, Ḥays wares from later centuries are crudely made.[22] The rather uneven quality of the excavated pieces suggests such a later date. Thus, the stepped sluiceway is an example of investment in the Wādī Zabīd in the post-Ottoman era. According to our local informant, Aḥmad ʿĪsā Aḥmad Hulaybī (the schoolteacher of Maḥall al-Shaykh), the man after whom the village was named is a certain Shaykh al-Ṣadīq al-Burjumī, who arrived in the area in 1557. The spillway fell into disuse as a result of the implementation of the Tihāma Development Project in the late 1970s.

20 See Keall 2001a, 42–43 and 48, fig. 3, especially nos. 8 and 13, for the typical fifteenth-century pottery with a white fabric carrying a design in blue paint; the considerably smaller types, fig. 3, nos. 9–12 and 14, date from the sixteenth century.

21 Keall 1991, 82–84; Keall 1992, 29–32; Keall 2001a, 42–43 and 49, fig. 4, especially nos. 2–4; and Ciuk and Keall 1996, 6.

22 See Keall 2001a, 43 and 49, fig. 4, especially nos. 16–18.

CASE STUDY 5

Water and Waste in the City of Zabīd

Besides what has been described in Case Study 3, other surprise discoveries were made during the Canadian Archaeological Mission's building-stabilization and conservation work in the compound of the citadel of Zabīd in the 1990s. They gave rise to a new direction in our archaeological investigations, with a focus on urban water delivery and waste removal. The excavations have afforded us detailed insights into the significance of these issues in the context of a medieval Islamic city, as well as into the engineering standards applied at the time.[1]

It is important to note that the citadel of Zabīd was built following the first Ottoman conquest of the city in 1539 and therefore no earlier than the mid-sixteenth century.[2] The construction incorporated considerably older buildings such as the Iskandariyya mosque-*madrasa*, and the citadel compound was set upon urban layers that date back to the Ziyadid period (820–1018).

1 The Five-Domed Building

In 1994, building-stabilization work began on a five-domed building (fig. 52, *D*) situated immediately to the west of the Iskandariyya mosque-*madrasa* inside the citadel (grid-system code: ZSE 26). The domes were poorly preserved and in danger of collapse (fig. 53). There was no living memory of what the structure had been used for, but one of the workmen made the suggestion that the five units had once been toilets. Debris clearance revealed that this had indeed been the case in the last formal stage of the building's life. Each of the domed chambers was fitted out as a toilet through the construction of a deep shaft that operated as the cesspit. A brick cone with an opening in its centre built over the top of the shaft provided the essential drain hole, as is the norm for a traditional Zabīd-style toilet.[3] At the time of construction, backfill was poured around the cone to ground level, and a floor was laid at a datum height of 106.39 m. The datum heights of the archaeological features described in this

1 An earlier version of the first part of this chapter was published in 2007 as "Water and Waste in Medieval Zabīd, Yemen."

2 Keall 1991, 79; the present-day layout also incorporates walls of a much more recent date (84–85).

3 The excavation of medieval toilet cesspits in Zabīd was first reported by Keall (1989a, 66).

 | DOI:10.1163/9789004387713_009

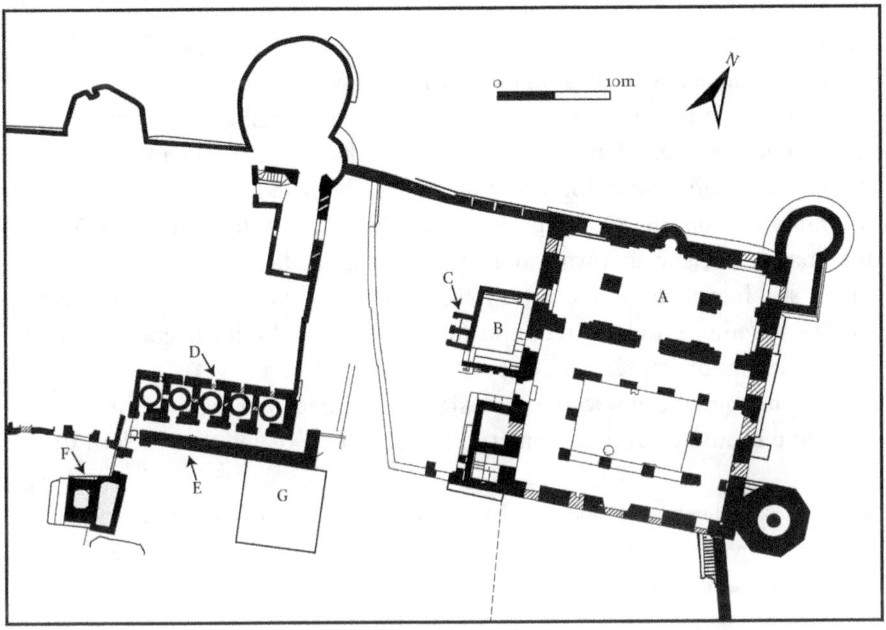

FIGURE 52 Plan of the northeast corner of the Zabīd citadel: the Iskandariyya mosque-*madrasa* (*A*) with its open ablution pool (*B*) and two private ablution cubicles (*C*); the five-domed building (*D*) and privacy wall (*E*); the adjacent well (*F*); and the excavation trench (*G*)

FIGURE 53 Five-domed building (with four domes visible) before restoration, with wellhead

chapter are listed in table 9, in chronological order from the earliest remains (phase 1) to the most recent (phase 7), which include the toilets.

To provide access to the toilets at these now elevated levels, steps leading up into the units were built. In four of the five chambers, the bricks in the wall above the original arched entrance were hacked out and a wooden beam inserted to form a new (higher) door lintel. The backfill included all manner of premodern debris, and hence it appears logical that the toilets were used by the imam's garrison that was housed in the Zabīd citadel until 1962; the toilets might still have been in use after 1962. Dereliction, however, gradually settled in. The wooden lintels rotted, and four of the five domes began to collapse, which is what prompted the 1994 stabilization work.

But the domed units were not designed as toilets. Removal of the toilet devices and clearance of their construction backfill revealed that each of the five

TABLE 9 Datum heights (above sea level) of archaeological features (ZSE 26)

Phase	Archaeological feature	Datum height (m)
1	bottom of earliest construction activity	96.45
2	– bottom surface inside wellhead holding tank	104.03
	– bottom of massive block of masonry, forming part of wellhead complex	101.05
3a	floor of earlier toilet	102.78
3b	floor of later toilet	103. 32
4a	earliest canal: bottom surface inside	103.80
4b	canal running along southern end of wellhead holding tank: bottom surface inside	104.06
4c	canal at northern end of wellhead complex: bottom surface inside	104.25
5a	closed canal: bottom surface inside	103.97
6a	– construction work surface for digging foundation trenches of five-domed building and brick privacy wall	103.70
	– original (lowest) base surface inside water trough at back of most easterly unit in five-domed building	104.95
6b	bottom of plaster on south face of privacy wall	
	– at east end of trench	104.46
	– at west end of trench	104.96
7	floor of toilet in middle unit of five-domed building	106.39

chambers was originally a bathing space, with a deep water trough at the back of the room (figs. 54 and 55), positioned at a datum height of 104.95 m (table 9, phase 6a).[4] Each trough was directly connected to the next one via an opening at the bottom of the wall. The water entered the units from the western end, flowing under gravity from an adjacent wellhead (fig. 52, *F*).

FIGURE 54 The most easterly unit of the five-domed building in its original state. It served as a private bathing space (*ḥammām*), with a deep water trough at the back of the room (50 cm scale)

4 As described below, the bathing facilities were renovated several times. The datum height given here refers to the original (lowest) base surface inside the water trough.

FIGURE 55 Detail of water trough (50 cm scale)

The five-domed building is in the immediate vicinity of the Iskandariyya mosque-*madrasa* complex and clearly provided ablution facilities (*ḥammāmāt*, sg. *ḥammām*) for the mosque. For privacy, in addition to each unit's doorway, a substantial wall was constructed (fig. 52, *E*). It ran parallel to the south side of the building before taking a right-angled bend to the north, creating a narrow corridor. This provision of closed-off, roofed chambers for bathing is a complete anomaly in the cultural record of Zabīd, where typically a mosque complex has an open ablution pool (*birka*), with one or two cubicles for private ablution partitioned off at the side but not roofed or totally secluded from outside view. The Iskandariyya mosque-*madrasa* ablution facilities serve as an example (fig. 52, *B* and *C*, and figs. 56 and 57).

The questions, then, of when these *ḥammāmāt* were built and by whom are of major concern for understanding how water was used in the urban context of Zabīd in general and, more specifically, in public ablution facilities. First, one may consider the brickwork of the structures. It contains whole bricks of varying shapes and sizes in addition to fragments of broken bricks (table 10), a clear indication of the employment of recycled bricks in the building's construction.[5] There is increasing evidence for brick recycling during the

5 Extensive brick-recycling activities in the Zabīd citadel were first described by Keall (1991, 80–81). See also Keall 2012, 129–131 and 137.

FIGURE 56
Empty ablution pool (*birka*) of the Iskandariyya mosque-*madrasa*, with bottom openings of two cubicles providing privacy for ablution (as seen from the entrance to the Iskandariyya mosque-*madrasa* courtyard)

early centuries of Zabīd, and the first textual reference can be found in Ibn al-Mujāwir's travel account. It relates events from the 1220s—that is, late Ayyubid times—when the ruler's residence, originally constructed with baked bricks during Ibn Ziyād's reign (820–859), was demolished, and the bricks, both whole and broken, were reused to build houses.[6] Brick recycling was practised extensively, possibly even exclusively, when the political and economic situation in the Tihama deteriorated after the fall of the Tahirids—that is, during the brief Mamluk interlude (1516–1517), the Mamluk-Lawandi military regime (1517–1539), and the first Ottoman occupation (1538–1636).

Second, the zone of transition from each square chamber to the circular dome of its roof is constructed with "saw-tooth pendentives" (fig. 58), a device in which the bricks in the individual corbelled rows "are set with one corner of each brick exposed, giving a 'saw-tooth' appearance."[7] This kind of pendentive replaced the earlier use of a squinch to span the corner and was adopted

6 Ibn al-Mujāwir/Smith, trans. 2008, 103–105.
7 Keall 2001a, 222.

FIGURE 57
Cubicles for private ablution (with entrance to the Iskandariyya mosque-*madrasa* courtyard)

TABLE 10 Zabīd brick typology (part 2)

Grid-system code	Archaeological feature	Brick size (cm)	Date	Basis for date	Remarks
ZSE 26	wellhead complex (phase 2)	varying shapes and sizes, whole and broken	post-11th c.	archaeological context: pottery from phase 1	bricks are recycled
ZSE 26	later toilet, brick cone (phase 3b)	16 × 16 × 3.5 to 18 × 20 × 4.5; and a few 13 × 13.5 × 3 (small-tablet type)	most probably 13th c.	pottery, including Chinese import	small-tablet type of bricks are recycled

TABLE 10 Zabīd brick typology (part 2) (cont.)

Grid-system code	Archaeological feature	Brick size (cm)	Date	Basis for date	Remarks
ZSE 26	canal at northern end of wellhead complex (phase 4c)	22 × 12 × 5–5.5	? (between phases 3b and 5a)	archaeological context	—
ZSE 26	closed canal (phase 5a)	varying shapes and sizes, whole and broken; cap bricks: 14 × 14.5 × 3.5 (small-tablet type)	most probably just before construction of the five-domed building and the privacy wall	archaeological context	bricks are recycled
ZSE 26	closed canal, repair (phase 5b)	17.5 × 26 × 5 to 17.5 × 26.5 × 4	most probably around the time of construction of the five-domed building and the privacy wall	archaeological context	—
ZSE 26	five-domed building and privacy wall (phase 6a)	varying shapes and sizes, whole and broken	around or after mid-16th c.	use of saw-tooth pendentives; (*terminus post quem* based on backfill between bathing units and privacy wall)	bricks are recycled
ZSE 36	upper brickwork carrying single lead pipe	20 × 20 × 4.5 to 20.5 × 20.5 × 4.5	ca. mid-14th c.	archaeological context and pottery, including Chinese import	—
ZSE 36	lower brickwork carrying three lead pipes	20 × 20 × 4.5 to 20.5 × 20.5 × 4.5	ca. mid-14th c.	archaeological context and pottery, including Chinese import	—

FIGURE 58 Saw-tooth pendentive supporting the dome in one of the units of the five-domed building

universally in Zabīd around the middle of the sixteenth century. The example of the Muṣṭafā Pasha al-Nashshār mosque-tomb complex in Zabīd has been cited as supporting evidence.[8] Muṣṭafā Pasha al-Nashshār served twice as the Ottoman governor of Yemen, first for a five-year term 1540–1545, and again for five to six months from late 1554 until his death in 1555.[9] The dome of the mosque—apparently sponsored during his first governorship—uses the squinch technique, while his adjoining tomb was built with saw-tooth pendentives.

Third, several replasterings of the floor surface of the Zabīd citadel's ablution facilities and even complete refloorings with brick subfloors attest to the fact that the *ḥammāmāt* continued to operate over a considerable span of time. Pottery recovered from different contexts gives an indication of dates associated with various building and renovation activities in the life of the bathing facilities.

For the construction of the *ḥammāmāt,* a foundation trench for the footings was dug from a well-defined work surface at a datum height of 103.70 m (table 9, phase 6a); the work surface contained pieces of plaster, ashes, and broken bricks. The bricks of the footings (baked brick in mud mortar) were laid

8 Ibid.

9 See Blackburn 1993, 721, for information on Muṣṭafā Pasha al-Nashshār.

approximately to the width of the trench. This left a rather ragged edge to the footings when exposed. As soon as the ground surface was reached, the walls were built straight up, with bricks laid to a vertical face.

The foundation trench for the footings of the privacy wall was dug from the same work surface. The wall itself was slightly set back from the footings. The change is clearly visible in the brick face, where a ledge indicates the top of the footings and the bricklaying changes from rather irregular to regular (both the wall and the footings are built from recycled bricks). For the footings, large quantities of mud mortar (*ṭīn*) were used. This resulted in lumps of mud mortar filling the void between the brickwork and the trench. Since the *ṭīn* is much denser than the dirt through which the foundation trench was dug, each could be identified very clearly during our stabilization work. Following the construction of the bathing units and the privacy wall, backfill was dumped into the space between them. This deep deposit, about 1 m thick, is composed of mixed building debris and contains masses of broken tiles, bricks, glazed earthenware water pipes, and large pottery sherds. The backfill was apparently derived from a considerably earlier building and did not necessarily originate from the same site. It does not date the *ḥammāmāt* construction work. However, as a whole, the material gives a *terminus post quem* of the early sixteenth century for the construction of the five-domed building. It should be pointed out that the deposit contained none of the fine-quality Ḥays pottery that is associated with the first Ottoman occupation.[10]

2 Water-Related Structures and Their Use before the *Ḥammāmāt*

In 1995, full archaeological excavations were conducted to the south of the five-domed building in order to determine the pattern of land use over time in this part of the Zabīd citadel (grid-system code: ZSE 26). The issue was explored further during the 2002 season of excavations and again 2005–2006. The deepening and widening trench has provided interesting information on urban water supply and waste removal, which were clearly important concerns over a period of several centuries. The particulars are presented here in chronological order starting with the earliest remains exposed (phase 1). See figure 59 for a plan of the excavation trench and the archaeological remains of the different phases. Figure 52 shows the location of the trench (*G*).

A deep sounding (fig. 59, *1*) enabled documentation of unequivocal evidence of repeated construction activity overlying virgin soil, which was reached at

10 For the characteristics of fine Ḥays pottery, see Case Study 4, section 6.

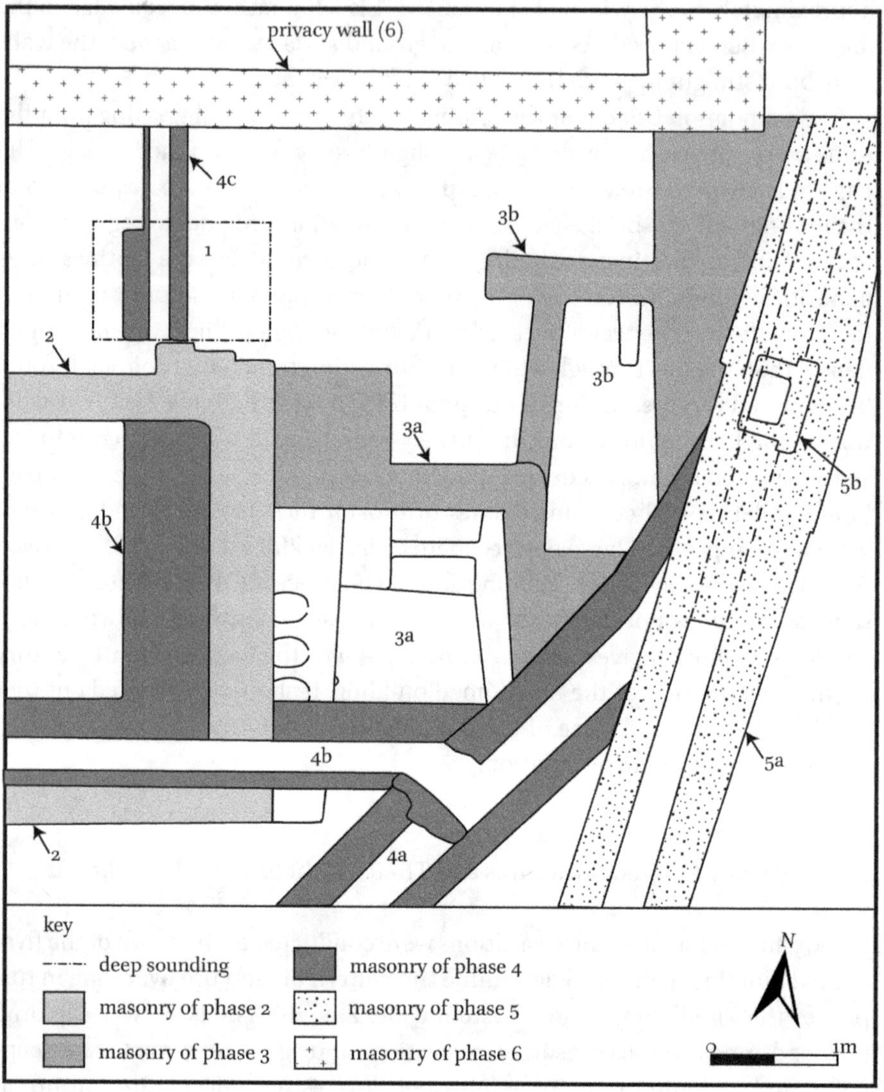

FIGURE 59 Plan of the excavation trench. The labels *1* to *6* indicate the different phases of construction

96.45 m. Since these layers are not precisely defined, they are labelled here loosely as "phase 1," and the detailed discussion of the findings starts with "phase 2."

2.1 *Phase 2: The Wellhead Complex*

The earliest extensive structure documented is a massive block of masonry that was erected to form the foundation of a holding tank (*ḥawḍ*, bottom surface at 104.03 m) for water drawn from what can be hypothesized as an adjoining well (fig. 59, *phase 2*).[11] The system would have been operated by a draught animal walking down a ramp and pulling up one bucket of water after another to fill the tank.[12] From here, the water could be used for drinking, washing, or small-scale irrigation. The massive masonry construction forming part of the wellhead complex was built with reused bricks of varying shapes and sizes (table 10), making precise dating difficult. The aforementioned deep sounding along its northern side exposed the bottom of the masonry at a datum height of 101.05 m. Material from beneath this level (phase 1) indicates that the block of masonry was built more recently than the eleventh century. From the evidence of phase 3, it is clear that the holding tank remained in use through the twelfth to thirteenth centuries.

2.2 *Phase 3: Two Toilets*

On the eastern side of the holding tank, set in a small open court (datum height: 102.78 m), is a well-built toilet (fig. 59, *phase 3a*). Next to it is a bench with plaster ledges that was designed to hold water jars (fig. 60). The distinctly sloping surface of the bench together with the openings of the horseshoe-shaped plaster ledges on the lower sides allowed water seeping from the jars to run off into the drain provided in the equally sloping floor of the toilet compound. The jars must have had a sufficiently delicate structure and no glaze to enable evaporation—and thus the cooling of drinking water. It may be suggested that, based on our understanding of disease transmission, the storage of drinking water next to a toilet reflects poor standards of sanitation. One has to remember, though, that the potential health hazards associated with human excrement were only understood around the middle of the nineteenth century.[13] For medieval Zabīd, the provision of cool drinking water reflects sophisticated standards of urban services.

---

11 Even though the well itself was not exposed, sufficient circumstantial evidence was found (and is presented below) to justify the assumption.

12 For a schematic drawing of one of these animal-operated water-lifting devices, see Christiansen-Weniger 1961, 28. A photograph of the traditional system in action, using three draught animals, can be found in Naval Intelligence Division 1946, fig. 29. Lewcock (1983, 302) describes the operation of this kind of well in the context of Ṣanʿāʾ.

13 People in the ancient world already had a clear notion of clean drinking water versus contaminated water. However, it was only in the mid-nineteenth century that, for the first time in medical history, a disease was described as spreading through water contaminated with human excrement. The evidence was provided by the early-Victorian physician John Snow, who had investigated the reoccurring outbreaks of cholera in Britain during the 1830s and 1840s. Snow presented his complete theory in the 1849 publication "On the Pathology and Mode of Communication of Cholera."

FIGURE 60 Earlier toilet (phase 3a) and bench, built against the eastern wall of the wellhead complex, with a row of stands for holding water jars to the left (50 cm scale)

The facilities of the toilet compound were repaired several times, with plaster being smoothed against the eastern wall of the wellhead. When the toilet fell into disuse because its cesspit shaft had filled up, the opening was carefully blocked with slabs of plaster and bricks. By comparison with present-day Zabīd, a toilet of this kind can be judged to have served an extended family of some thirty people for a generation. How long a public toilet would have lasted is a moot point. Once the toilet was abandoned, trash started to accumulate in the now-derelict space. The layer of dirt overlying the floor contained a great amount of animal bones, large sherds of pottery (mostly from large, delicate, and unglazed water jars that would have been used for keeping drinking water cool and would have fitted in the horseshoe-shaped plaster ledges described above),[14] and even complete pots. This pottery is datable to the eleventh to thirteenth centuries.[15]

---

14 For examples of these jars, see Ciuk and Keall 1996, plates 95/15 and 95/16.

15 One specifically identifiable sherd of Chinese import was a fragment of Southern Sung celadon with a characteristic fish-motif decoration. Production of Southern Sung celadon started in 1127 and continued beyond the end of the Sung Dynasty (960–1279) into the Yuan period (1279–1368), perhaps as late as the early fourteenth century. For two pieces similar to the one excavated in Zabīd, see National Palace Museum Taipei 1974, plates 25

FIGURE 61 Excavated brick cone of later toilet (phase 3b; 50 cm scale)

At some point, a new and slightly less elaborate toilet was constructed in an adjoining space (fig. 59, *phase 3b*), with its floor level approximately 0.5 m higher (datum height: 103.32 m). Excavation on the exterior of its cesspit exposed the traditional construction pattern of a Zabīd-style toilet, with a brick cone built above the deep shaft dug for the excrement (fig. 61; see table 10 for the brick sizes). In the layers of dirt around the lower brick lining of the cone, a circular zone with a distinct yellow stain characteristic of urine was noticeable; bricks and potsherds showed the same yellow stain.

2.3 *Phase 4: Three Open Canals*

The entire toilet compound was eventually abandoned. Some time afterwards, the space was remodelled and put to a new use still involving water. A series of devices were constructed over time, each one slightly different than the others, and all designed to deliver water. First, a canal (fig. 59, *phase 4a*) was built, cutting across the space and running in a northeasterly direction towards the Iskandariyya mosque-*madrasa* complex, presumably to an open ablution pool in the traditional manner of Zabīd (figs. 52, *B*, and 56).[16] To construct the canal,

---

and 26. I am indebted to Patty Proctor, curator emerita of Chinese Ceramics, Department of World Cultures, Royal Ontario Museum, Toronto, for having provided this information.

16 For pools of this kind in Zabīd, see the observations made by Ibn al-Daybaʿ that are cited below in section 3.

a trench was dug from the existing surface. The bricks for the foundations were laid in mud mortar, and the two walls were built up against the sides of the trench, leaving a 40 cm wide canal for the water (datum height: 103.80 m). Finally, the inner faces were made watertight by lining them with lime plaster. Since the canal was built running in open ground, with its shoulders level with the existing surface, the outside walls could not be plastered and therefore appeared uneven and unfinished following excavation.

The canal was later modified when its western wall was chopped through and the canal itself blocked just before this spot in order to accommodate a new intake. At the same time, the wellhead holding tank was remodelled. This involved filling in part of the basin with bricks laid in lime mortar, thereby reducing the size of the tank while creating a canal (datum height: 104.06 m) running along the southern end of the original structure (fig. 59, *phase 4b*). A connecting piece had to be built between the wellhead complex and the existing canal, but the destination of the water flow remained unchanged.[17]

Remains of a third canal (phase 4c) can be found at the northern end of the massive wellhead complex. It ran south–north (datum height: 104.25 m) and was demolished when the privacy wall was built, possibly to make bricks available.[18] See table 10 for the brick sizes of the phase 4c canal. Other plastered surfaces in the wellhead indicate several remodellings of the system. While it is not possible to identify each individual modification, it can be concluded from the building's remains that all of them were related to water, and that the overall structure was used in some fashion for a considerable length of time.

2.4 *Phase 5: The Closed Canal*

The most recent water-delivery device in the trench is a closed canal (phase 5a). Its sides were built up two bricks wide, leaving a width on the inside of 35 cm (datum height: 103.97 m). The canal was roofed by an arch consisting of bricks laid in lime mortar, with its cap formed by bricks set vertically on their point (fig. 62). The exterior was coated with lime plaster, leaving only the top points of the brick cap exposed in a row. In order to make the canal watertight, its inner faces were lined with lime plaster. A wide range of recycled bricks

17 If one assumes that the canal was now fed from the well, the walkway of the draught animal would have been on the western side of the well. See also n. 20.

18 This example also explains a sometimes confusing characteristic of walls built with recycled bricks, such as the privacy wall including its footings: even though the wall was constructed with bricks laid in mud mortar, a number of bricks have lime mortar attached, derived from the original building from which they were recycled.

were used in the construction. Both the straight sides of the canal and the smoothed plaster on the outside indicate that it was built above ground. The canal runs straight across the southeastern part of the trench and then lines up with the earlier open canal (phase 4a), which it uses as its foundation (figs. 59 and 63). It obviously delivered water to the same destination—an open ablution pool next to the Iskandariyya mosque-*madrasa*—only at a slightly higher elevation.

The closed canal was most probably built just prior to the five-domed ablution facilities and the privacy wall. Its northern part was either damaged in the process of the *ḥammāmāt* construction or remodelled at a later point. As part of the reconstruction, an inspection chamber, or manhole, with a rectangular opening of 25 × 65 cm, was set above the canal to allow access for maintenance work such as cleaning and repairs (figs. 59, *phase 5b*, and 62). The rebuilt part of the canal is not nearly as regular and straight as the original part. The new cap was corbelled, and its flat top covered with a thick layer of lime plaster. The repair work was carried out with bricks of a totally different shape and size than the rest of the canal, measuring 17.5 × 26 × 5 cm to 17.5 × 26.5 × 4 cm (table 10).

FIGURE 62 Closed canal (phase 5a), with the rectangular opening of the inspection chamber (phase 5b; 50 cm scale), as viewed from the southern end of the trench. Note the downward slope (towards the east) at the bottom of the plaster facing on the privacy wall (phase 6)

FIGURE 63 Overall view of the excavation trench ZSE 26, with remains of the privacy wall at the back

2.5 *Phase 6: The Five-Domed* Ḥammāmāt *and the Privacy Wall*

As described above, the domed ablution facilities and the privacy wall were built at the same time. The layers of dirt accumulated against the outside of the privacy wall provide details about the activities on its southern side following construction of the five bathing units. At some time during its life, the southern face of the wall was plastered (phase 6b), with the plaster ending about 0.5 m above the footings. Most probably, this was done after the height of the wall was increased. While a distinct change in the brickwork is not discernible because of the presence of the plaster facing, construction activity is indicated in a building-debris layer that consists of pieces of brick, pebbles, and pottery sherds.[19] This debris is overlain by a whitish zone of plaster that lines up against the bottom of the plaster facing on the wall. Both the plaster zone and the bottom of the plaster facing show a noticeable slope down towards the east (table 9, *phase 6b*, and fig. 62). Within the trench, the plaster zone also slopes distinctly from the brick privacy wall towards the south. It extends into the southern half of the trench as a dense, compacted mud surface. The

19 The quality of the bricks used in the upper layers of the privacy wall—above the plaster facing—is quite different from the lower layers just overlying the footings. While the entire wall was built with recycled bricks, its upper part contains predominantly small brick fragments. This also explains why the top of the wall is in such a poor state of preservation.

pottery that was recovered consisted of tiny pieces, as opposed to the large fragments found right next to the privacy wall. Taken together, this is clearly evidence for the existence of a well ramp, to allow a draught animal to pull up buckets of water from the well while walking down the ramp.[20] As for the date of the wall plastering, it is notable that the layers described here are the lowest dirt layers in the trench from which fine-quality Ḥays pottery sherds were excavated, and this implies that the plastering cannot have occurred before the first Ottoman occupation.

More recently, the top of the privacy wall was modified to carry a succession of water conduits that were fed from the existing well (fig. 52, *F*) and delivered water to the open ablution pool of the Iskandariyya mosque-*madrasa* located to the northeast.

3 Historical Context

Not surprisingly, it is Ibn al-Daybaʿ who provides the most significant information in this context. His work includes references to the urban life in his hometown, Zabīd. In his *Bughyat al-mustafīd*, Ibn al-Daybaʿ frequently mentions mosque ablution pools. During the period of Rasulid rule (1228–1454), sponsoring mosque ablution pools seems to have been an established practice, as exemplified by Jihat Farḥān (d. 1432/3), the wife of the seventh Rasulid sultan, al-Ashraf Ismāʿīl I (r. 1377–1400). She not only sponsored the Farḥāniyya *madrasa* in Zabīd but also the pool (*birka*) of the Ashāʿir mosque around the year 1413.[21] The Rasulid sultan al-Ashraf Ismāʿīl III (r. 1438–1441) ordered the construction of a large, beautiful pool for the Grand Mosque (*jāmiʿ*) of Zabīd.[22]

The third Tahirid sultan, al-Manṣūr ʿAbd al-Wahhāb (r. 1478–1489), rebuilt the Ashāʿir mosque, which involved remodelling of its ablution-pool compound.[23] As well, he built the small ablution pool of the Grand Mosque.[24] His son al-Ẓāfir ʿĀmir II (r. 1489–1517) had a modification made to the ablution facilities of the Manṣūriyya *madrasa* built by his father. The neighbours had

20 It should be noted that the two consecutive *open* canals cutting across the southeastern part of the trench mean that at the time of their functioning, no ramp for the draught animal can have existed on the eastern side of the well.

21 Ibn al-Daybaʿ/Chelhod, ed. 1983, 108; and Sadek 1989, 122 and 124. On archaeological excavations in the Ashāʿir mosque, including a probe through a derelict ablution pool that might even be the one sponsored by Jihat Farḥān, see Keall 2012, 132–135.

22 Ibn al-Daybaʿ/Chelhod, ed. 1983, 115.

23 Ibid., 112–113.

24 Ibid., 172.

filed a complaint about inappropriate disposal of the waste water from the washrooms. Sulṭān al-Ẓāfir ʿĀmir II ordered a new drain to be built to channel the water out of the city.[25]

The Grand Mosque already had two ablution pools (one built by the Rasulid sultan al-Ashraf Ismāʿīl III; the other, by the Tahirid sultan al-Manṣūr ʿAbd al-Wahhāb) when Sulṭān al-Ẓāfir ʿĀmir II ordered the renovation of the mosque in 1492. During the course of the work, the master builder (*muʿallim*) suggested a third ablution pool be built on an adjacent piece of land that was separated from the mosque by a street. This was rejected on the grounds that "the path of the Muslims" should not be changed.[26] However, when the workers excavated trenches to expose the mosque foundations, they discovered building remains that led across the street to the aforementioned property. These were eventually identified as remnants of a large *birka*, surrounded by partitioned bathing spaces for private ablution (*mughtasalāt*, sg. *mughtasala*) and public toilets (*kunuf*, sg. *kanīf*). Needless to say, the people of Zabīd understood the significance of these findings, and the new *birka* was built with the same layout after all. Ibn al-Daybaʿ suggests that the original pool had most likely been constructed by the prolific Ziyadid ruler (and builder) Ibn Salāma (r. 983–1012), whom he credits with having founded the Grand Mosque.[27]

The various structures fed by wells tell us about the wells' importance for supplying the urban water system. This was already a fact by late Ayyubid times. Ibn al-Mujāwir, who visited Zabīd towards the end of Ayyubid rule, stated: "The water of the town is from wells."[28]

---

25 Ibid., 196.

26 Ibid., 82.

27 Ibid., 80–82. Interestingly, when describing the problems related to the construction of the third ablution pool (81–82), Ibn al-Daybaʿ reports the events as having taken place during renovation of a mosque that he calls Masjid al-Manākh, a name not otherwise mentioned in the historical sources. From the description of this mosque and the date of its renovation, one can only conclude that, in fact, Ibn al-Daybaʿ is referring to the Grand Mosque. He specifically points out that the two existing ablution pools had been built by the Rasulid sultan al-Ashraf Ismāʿīl III and the Tahirid sultan al-Manṣūr ʿAbd al-Wahhāb, both of whom are credited elsewhere in the book with having sponsored ablution pools in the Grand Mosque, as outlined above. In addition, Ibn al-Daybaʿ gives the date 1492 for the renovation of the Grand Mosque (*jāmiʿ*) (186), the same date that he gives for the Masjid al-Manākh renovation (80–81). See Keall 1984, 53, for two dated inscriptions in the Grand Mosque marking its 1492 renovation under al-Ẓāfir ʿĀmir II. Porter (1992, 209–211 and 217–220) provides additional evidence for the identification of Masjid al-Manākh with the Grand Mosque.

28 Ibn al-Mujāwir/Smith, trans. 2008, 110.

4 Religious Implications

It is one of the universal principles in the Islamic world that bathing facilities must be provided for ritual ablutions, ceremonial washings not concerned with hygiene. A Muslim can neither perform a valid prayer nor touch a copy of the Koran unless he or she is in the state of ritual purity (*ṭahāra*). Islamic law distinguishes between "major" and "minor" ritual impurity.[29]

A state of minor ritual impurity (*ḥadath*) stems from, for instance, contact with a substance that is regarded as unclean, such as wine, dogs, or excrement. To regain purity before prayer, a Muslim must perform the minor ritual ablution (*wuḍūʾ*) as given in the Koran, by washing the face, hands, and feet. As a rule, mosques provide washing facilities for performing minor ritual ablutions.

Major ritual impurity (*janāba*) ensues from sexual intercourse and, for women, also from menstruation and childbirth. It requires the major ritual ablution (*ghusl*) and involves a thorough washing of every part of the body including the hair. Major ritual ablution is also recommended on other occasions: for example, before attending the congregational Friday prayer. The need to perform major ritual ablutions before prayer explains the partitioned bathing spaces providing privacy.

5 Conclusion

From Ibn al-Daybaʿ's work we get a good sense of the significance of water and waste in the religious context of the city of Zabīd. However, while engineering schemes are mentioned, information regarding the technical and functional details is lacking. This is a general rule in the ancient and medieval history of technology: the texts usually give no more than the historical framework within which the structures may be placed, and they rarely provide technical descriptions or explanations of how they functioned. For these, one must resort to archaeological excavation as an investigative tool. It is the only approach that exposes the engineering schemes and affords detailed documentation and analysis of their technical traits. The best sources of information on water and waste are the canals, toilets, and bathing facilities themselves.

---

29 The general rules for purity and ablution given here follow Gibb and Kramers (1953) 1974, and Glassé 1989. It should, nevertheless, be pointed out that these rules vary according to the different branches of Islam. The same is true of the regulations for the minimum amount of water required for purification.

On the basis of the archaeological evidence, it can safely be stated that the area under investigation in the Zabīd citadel (ZSE 26) has been consistently used for water delivery and waste disposal for the past thousand years. The building activity and many remodellings are not only preserved in the water-related structures, but they are also well documented in the layers of dirt and accumulations of building debris excavated from the trench. While the technical details provide information about the engineering skills of the builders, the structures more importantly reflect upon the services of water and waste management that were available to ordinary people in Zabīd, and thus upon the quality of life in a medieval Islamic city.

6 Postscript

In the Zabīd citadel, some 100 m to the south of the five-domed building where the masonry canals were exposed, a second trench was first opened in 1989 and expanded in several field seasons (grid-system code: ZSE 36). During the excavations in 1990, a continuous stretch of brickwork (baked brick in lime mortar) was discovered (fig. 64, *A*). It was designed to carry water in an encased lead pipe.[30] Subsequent deepening of the trench has shown that there is, in fact, a second stretch of brickwork (fig. 64, *B*) carrying three lead pipes running below *A*; the two are not aligned.

All four pipes, each with an external diameter of 3–4 cm, were formed from lead sheets by bending the sheet around a wooden core that was then withdrawn. Pipes with overlapping edges were the result. Instead of being welded to make them watertight, they were covered in a layer of mortar up to 2 cm thick and wrapped in sheets of cloth to keep the mortar in place.[31] The wrapped units were then laid in a thick deposit of mortar in the space left in the brickwork for the pipe bed (fig. 65).[32] While the cloth fibre has disintegrated, impressions of the fabric are visible in the mortar (fig. 66). The space

30 Keall 1991, 81 and 90, fig. 5.

31 The fabric was analyzed by Gillian Vogelsang-Eastwood, Textile Research Centre, Leiden, The Netherlands, who identified it as being (most likely) linen.

32 Mortar as a construction material is defined as "any material used in a plastic state which can be trowelled, and becomes hard in place, and which is utilized for bedding and jointing" (Cowper 1927, 51). Mortar should not be confused with plaster, which is "any material used in a plastic state to form a durable finishing coat to the surfaces of walls and ceilings" (Cowper 1927, 29). As pointed out by Lucas (1944, 62), the respective term "depends upon the use to which the material is put and not upon its composition." In short, the distinction is between bedding and jointing on the one hand, and coating and lining on the other.

FIGURE 64 Overall view of the excavation trench ZSE 36, with two stretches of brickwork: the upper one (*A*) carrying one lead pipe and the lower one (*B*) three lead pipes

immediately surrounding the wrapped lead pipes was filled with a loose mixture of building debris, broken pottery, and powdery dirt before it was closed by baked bricks in lime mortar. Both brick structures carrying the lead pipes were built with bricks of the same size (table 10). Their dating, based on the archaeological context and the pottery associated with it, was explained in detail in a publication following the 1990 field season.[33] Needless to say, continuing excavations both in Zabīd and elsewhere have led to some revisions of the Zabīd pottery typology.[34] On this basis, it is most likely that the pipes were constructed around the mid-fourteenth century.

While systematic investigation of the engineering standards and technical traits of urban water-supply systems is only just beginning for the medieval Islamic city, they are reasonably well documented in classical antiquity. With regard to water-conveying systems, the Roman architect and engineer

33 Keall 1991, 80–83; the two pipe beds were constructed at the same time as building A-v, and the material recovered from the layer beneath the street running along its northern side provides the important clues.

34 See Keall 2012, 131 132 and n. 2.

FIGURE 65 Lead pipe in lower stretch of brickwork (5 cm scale)

FIGURE 66 Detail of the lead pipe, with an impression of the fabric wrap visible in the mortar

Vitruvius (1st c. BCE) states in his *Ten Books on Architecture*: "There are three methods of conducting water, in channels through masonry conduits, or in lead pipes, or in pipes of baked clay."[35] The two trenches in the Zabīd citadel provide archaeological evidence for the first and the second method in medieval Yemen, while various trenches to the east and to the north of the city that were described in Case Study 4 (sections 2–4) attest to the third method.

35 Vitruvius/Morgan, trans. 1960, 8.6.1. See the Introduction to this book, section 5, on "canal" vs. "channel."

PART III

Water and Settlement on the Coastal Plain of Southern Yemen: The Example of Ghayl Bā Wazīr

The title intentionally echoes John C. Wilkinson's 1977 work *Water and Tribal Settlement in South East Arabia: A Study of the* Aflāj *of Oman.*

Introduction to Part III

The examples of Mārib and Zabīd have shown the complete dependence of both settlements on their natural environment. The full scope of how local environmental factors—such as the available water resources and a region's geology—shape the approaches taken in hydraulic engineering and water management can be observed in a unique way in the Ghayl Bā Wazīr area. Ghayl Bā Wazīr is a town in the southeastern Yemeni province of Ḥaḍramawt, some 15 km inland from the Gulf of Aden coast between today's provincial capital of al-Mukallā and the port of al-Shiḥr (see fig. 67). Al-Shiḥr and its trade relations with India are mentioned in the tax register of the fourth Rasulid sultan, al-Mu'ayyad Dāwūd (r. 1296–1321), compiled at the end of the thirteenth century.[1] Ghayl Bā Wazīr is not listed in any of the Rasulid manuscripts cited in previous chapters.

The term *ghayl* as part of the town's name points towards the fact that the settlement's existence is intimately connected with the management of underground water flow, since in colloquial Yemeni usage, in an urban context *ghayl*, pl. *ghuyūl*, has the meaning of "an artificial often partly subterranean water-channel."[2] It was a *ghayl* that gave rise to a prosperous settlement based on irrigation agriculture.

From the start, field research in Ghayl Bā Wazīr depended on teamwork. Following a brief visit in 1998, two field surveys of the channelled landscape of Ghayl Bā Wazīr (January 2001 and December 2003–January 2004) provided us with the opportunity to investigate one of the underground water systems of Yemen, most importantly one that was then still partly working. As in the Wādī Zabīd, in Ghayl Bā Wazīr we were able to observe first-hand a traditional system of water engineering and management, and this on-site experience helped us to identify the technical principles and patterns of water use that supported a settlement and its agricultural economy. Field research encompassed Ghayl Bā Wazīr and its immediate hinterland. As it turned out, the territory where water systems had been set in place considerably exceeds the study area. In fact, from al-Mukallā, some 40 km to the southwest of Ghayl Bā Wazīr, to al-Ḥāmī, 60 km to the east (see fig. 67), there are numerous sites where underground stream-flow management has been practiced, though none of the systems are

1 Jāzim 2008, 129 and 136.

2 Serjeant, Costa, and Lewcock 1983, 19. Please note that *ghayl* can have different meanings. See Part II, Introduction, section 2, for the meanings "base flow" and "perennial stream." See the Introduction to this book, section 5, for "canal" vs. "channel."

 | DOI:10.1163/9789004387713_010

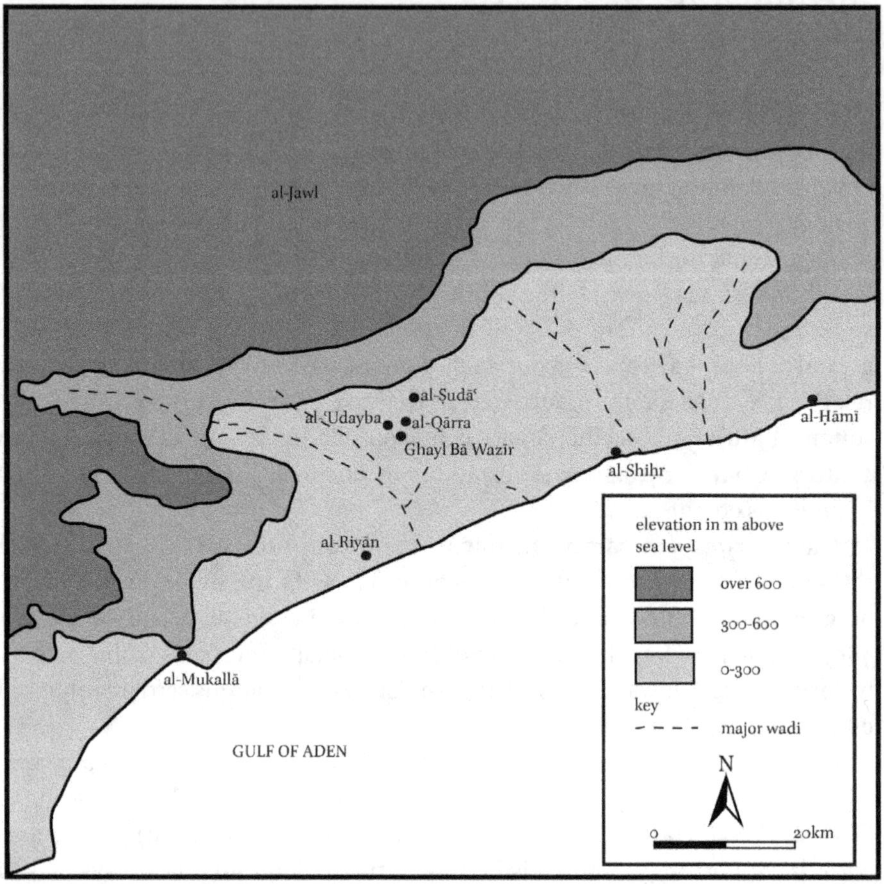

FIGURE 67 Ghayl Bā Wazīr's location on the coastal plain of southern Yemen

as complex or as technically challenging as those of Ghayl Bā Wazīr.[3] For a map of the study area, see fig. 68.

I am particularly grateful to individuals from Ghayl Bā Wazīr and the greater Ghayl Bā Wazīr area for sharing their knowledge and expertise with us. Their contribution to the following case study is invaluable because we have not been able to locate any written historical sources that deal with the technical details of the underground canal system. As outlined in the context of Zabīd, this is not

3 Many of the water sources are hot springs, some of which are sulphurous. They were first mentioned by Bent (1894, 317), and more detail can be found in Naval Intelligence Division 1946, 150. See also Lightfoot 2000a, figs. 3 and 8. To this day, the hot sulphur springs attract visitors from the area and beyond because of their healing effects on rheumatic, skin, and other diseases.

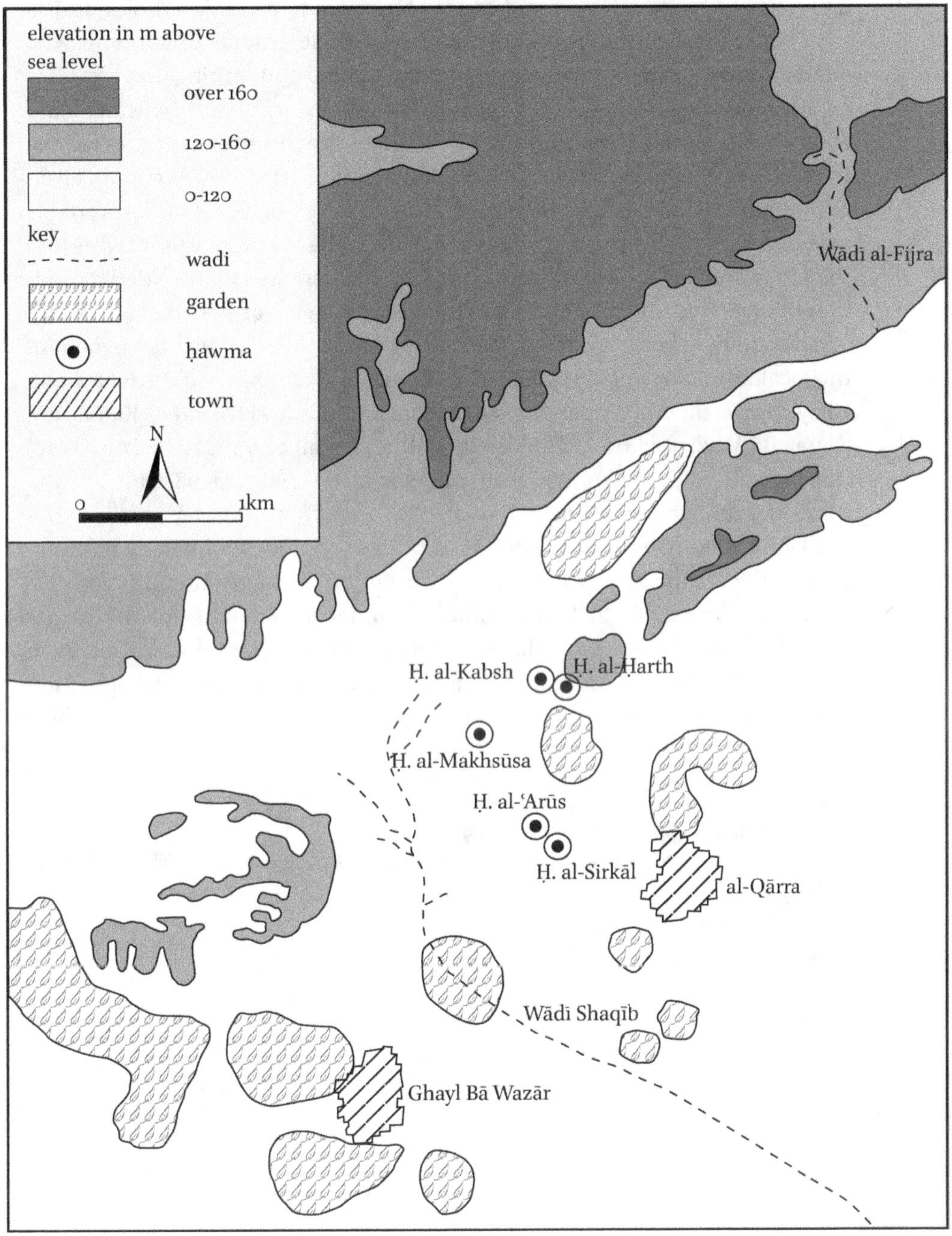

FIGURE 68 Map of the Ghayl Bā Wazīr study area

an unusual situation. We were therefore fortunate to meet ʿAwaḍ Bā Ḥamīsh, who had worked in his youth as a digger of underground canals, and who clarified the steps that are involved in the planning and digging processes. He greatly helped in making sense of some of the engineered features. Muqaddam Sālim, a former supervisor of one of the underground canals, addressed the managerial side, in particular the water-allocation practices. Accounts given by Sālim Bā Qaḥayzal, from the Environment Protection Council, Ṣanʿāʾ, of his experience growing up as a farmer's son in Ghayl Bā Wazīr gave us insights into the operation of the system, which no longer functions as a whole. His coauthored 1996 field study of the surviving water systems of the Ghayl Bā Wazīr area formed the basis for many of our discussions.[4] We also had the benefit of the local knowledge of ʿUmar al-ʿAfārī, director of the Office of Agriculture in Ghayl Bā Wazīr; ʿAbbās Bā Wazīr, a descendent of Shaykh ʿAbd al-Raḥīm ibn ʿUmar Bā Wazīr (whose significance will be explained in Case Study 7, section 7); and ʿAbd al-ʿAzīz Bin ʿAqīl, director of the Ḥaḍramawt Office of the General Organization of Antiquities and Museums (GOAM), al-Mukallā. Other local informants whom we encountered by chance provided valuable information on the complex system of water distribution and allocation. We consulted the (few) available historical sources in order to contextualize the results from the field surveys. As well, during fieldwork, it was important for us to record the local terms and expressions used in the construction and operation of underground canals.[5]

4 See Bā Qaḥayzal, Saʿīd, and Ibn Ghūth 1996.

5 Preliminary observations on Ghayl Bā Wazīr were published in Hehmeyer, Keall, and Rahimi 2002, 83–97.

CASE STUDY 6

Tapping Underground Water: The *Maʿyān* System of Ghayl Bā Wazīr

1 The Geological Context

For a general impression of the geology of the study area, we may refer to H. O. Little's report on the geological history of the Mukallā hinterland. Little visited Ghayl Bā Wazīr in the spring of 1920 as part of a survey that he had been invited to carry out in southern Yemen. He mentions the massive gypsum shield immediately to the north of the town, suggests that it has its origins in the Miocene, and describes the sequence of geological events that created it. Tectonic uplift resulted in the folding of a ridge parallel to the coast. The gypsum would have formed as an evaporite deposit in shallow basins or lagoons, before eventually becoming exposed when the lagoons dried up. Subsequent faulting occurred in the Pleiocene, which helps explain why some of the gypsum massifs in the region of Ghayl Bā Wazīr are elevated as much as 100 m above the more or less flat shield behind the town. The landscape began to take on its current shape when torrential rains during that epoch eroded the drainage systems that we can observe today.[1]

The main part of our study area lies on this gypsum shield, technically a "karst," which extends from the northern edge of the town of Ghayl Bā Wazīr as far as the foot of the plateau of al-Jawl (see figs. 67 and 68). The term "karst" is the Germanized form of an ancient Slovenian word meaning "bare stony ground."[2] Surface weathering of the exposed rock has resulted in a distinctively hard, pitted crust. Beneath, the calcium-rich rock is fractured by numerous joints. It is also moderately water soluble. Water seeping along the subsurface joints steadily dissolves the rock, which ultimately leads to the development of a subterranean flow network. Typical karst environments are dry and rocky, with very little surface vegetation, and have underground drainage rather than surface streams.[3]

Over time, the subterranean water flow and the solution processes can enlarge the joints and create caverns that hold considerable volumes of water

1 Little 1925, 100 and 117–120.

2 Bloom 1978, 136; and Chorley, Schumm, and Sugden 1985, 181.

3 Bloom 1978, 136–148.

 DOI:10.1163/9789004387713_011

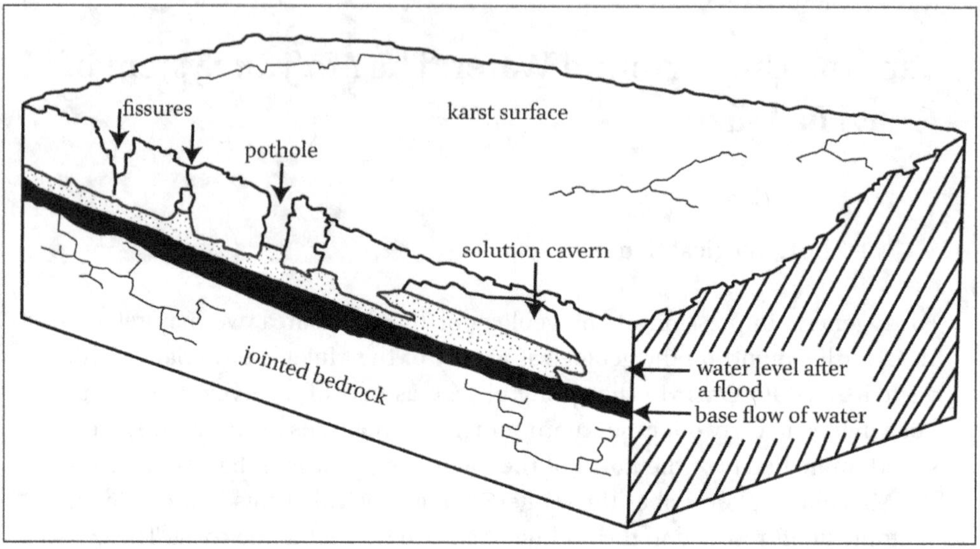

FIGURE 69 Typical water movement in a karst environment

(fig. 69). The rock ceiling of such a "solution cavern" may eventually become thin and collapse, in which case an abrupt breach with almost vertical sides or even overhanging cliffs appears at the surface.[4] The technical term is a "collapse sink" or a "collapse doline"; in the dialect of the Ghayl Bā Wazīr area, it is called a *ḥawma*, pl. *ḥuwam* (fig. 70).[5] The abruptness of the opening is reflected in local names such as Ḥawmat al-ʿArūs (literally, "the Collapse Sink of the Bride") and Ḥawmat al-Kabsh (literally, "the Collapse Sink of the Ram"). Both allude to the unsuspecting beings that, according to legend, fell down the holes in the dark and drowned.[6] The most recent collapse of a solution cavern in the area happened in 1976, when the small al-Ḥawma al-Jadīda (literally, "the New Collapse Sink") opened up; al-Ḥawma al-Jadīda is also called Ḥawmat Saʿd after the man who made its discovery by falling into it.

It is important to recognize that the distinctive collapse sinks are directly related to the karst landform, and that the processes involved in their formation result from the underground movement of water. The water held in the saturated bedrock, the solution caverns, and the collapse sinks is interconnected; the water level in the *ḥuwam* coincides with the water table and thus can

4 Ibid., 149–151 and fig. 7–10*b*; and Chorley, Schumm, and Sugden 1985, 186–187 and fig. 8.8*B*.

5 See also Serjeant 1964, 58. A *ḥawma* is filled with water year-round and distinguished from a rain-fed and therefore periodically dry waterhole that is locally referred to as a *karīf*.

6 A version of the story of Ḥawmat al-ʿArūs can be found in Serjeant 1964, 58.

FIGURE 70 Collapse sink (*ḥawma*) north of Ghayl Bā Wazīr (Ḥawmat al-Sirkāl)

fluctuate during the year and also as a result of particularly dry or wet years. The water resources are replenished by seasonal precipitation on the plateau of al-Jawl to the north of Ghayl Bā Wazīr, beyond our study area.

In the past, the collapse sinks were the most obvious source of water in a bleak and arid environment. Even though the water in the *ḥuwam* is exposed to the open air, it is ultimately groundwater. It can be used for irrigation agriculture as well as domestic purposes. However, making use of the water was not without difficulties. The water level in the collapse sinks is at a considerable depth below ground surface, and extracting it at the *ḥuwam*, especially in large volumes, would have required mechanical lifting devices. Moreover, the geological reality in the immediate vicinity of the water-filled collapse sinks means that there is no soil, just bare gypsum rock. Soil for plant cultivation is only found several kilometres to the south. The challenge, then, was to find an engineering approach to lead the water by gravity flow from the *ḥuwam* to where it was needed: the fields some distance away and the human settlement.

2 Connecting Underground Canals to Collapse Sinks: The Engineering Principles of the *Maʿyān* System

The obvious technical solution involved tunnelling from a point close to the intended destination through the bedrock all the way back to the *ḥawma* in order to tap the water. The total distance could amount to several kilometres,

FIGURE 71 An unfinished shaft (20 cm scale)

though. How to create the correct gradient for the flow of water? How to get the labourers to their work site in the morning? How to remove the debris excavated underground, and how to provide ventilation? These were serious engineering and logistical questions that required close attention.

First, the terrain from the *ḥawma* to the destination of the water had to be surveyed in order to assess the lie of the land and determine the course of the underground canal and its gradient. This was followed by cutting vertical shafts, like separate wells in a row, from the surface down into the rock along the route of the projected canal (fig. 71).

An experienced engineer would oversee the task of establishing the correct depth of each of the shafts. The completed shafts—called *nuqab*, sg. *naqba*—were connected through a gently sloping gallery at the bottom, referred to locally as a *shaṭṭ*, pl. *shuṭūṭ*, which needed to be sufficiently high and wide to accommodate the diggers as well as future maintenance workers. See figure 72 for a schematic drawing.

Besides ventilation and light, the shafts provided access and facilitated removal of the excavated rock. A man would stand at the surface and haul up buckets of debris that could be disposed of in three different ways. We find long rows of large circular spoil heaps around the shaft openings, with a diameter of some 5 m and up to more than 1 m high. To prevent the spoil from slipping back down the shaft, some of these spoil craters have carefully constructed retaining walls of broken rock pieces laid to a face. Alternatively, spoil

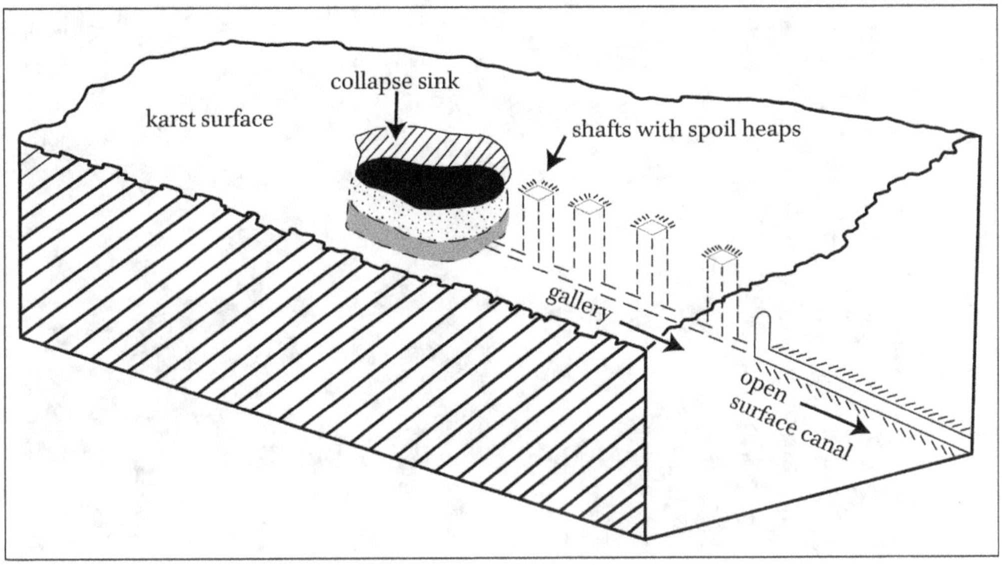

FIGURE 72 A partly underground canal (*maʿyān*) drawing water from a collapse sink (*ḥawma*)

banks were piled up parallel to one or both sides of the gallery below. Parts of the banks were constructed with the same kind of retaining walls (fig. 73). Finally, where the excavations produced good quality gypsum, it was removed for use in the building industry. Therefore, entire sections of some canals are conspicuous for the complete absence of spoil.

Digging started at the lower end of the underground canal and moved back towards the *ḥawma* in order not to flood the gallery prematurely; excavating the canal and finishing its sides was only possible in dry surroundings. The final act of connecting the gallery to the collapse sink was done by an experienced digger who had to cut the rock and dive immediately through the opening, with water now gushing in. His coworkers would throw him a lifeline and pull him up from the *ḥawma* to dry ground.

The rectangular shaft openings measure between 0.5 × 0.7 m and 0.7 × 1.05 m; close to the collapse sinks the shafts can be more than 13 m deep. Footholds to enable climbing up and down are usually provided on two opposite faces, either as small protruding ledges or as cavities cut into them (fig. 74). There are many examples where the tool-marks are pristinely preserved on the shaft faces. Their curving lines result from the rotation of the artisan's elbow as he pounded the chisel with a hammer (fig. 75). In the shaft corners, where the worker's arm was not free to swing in the same way, the cut marks are short and vertical.

FIGURE 73 Spoil bank with retaining wall

FIGURE 74 Footholds on a shaft face (30 cm scale)

The shafts were dug mostly at intervals of 5–10 m (8–9 m on average) along the length of the underground canal, but there can be as little as 3 m from one centre to another. At first sight, the spacing seems unnecessarily close. However, it can be argued that since engineering a new canal was a costly enterprise, digging it as quickly as possible with a large number of labourers would

FIGURE 75 Tool-marks on a shaft face (8 cm scale)

ensure fast return on investment. This required numerous cleanout shafts to accommodate the individual teams and avoid congestion in the gallery.[7] The landscape to the north of Ghayl Bā Wazīr is dotted with thousands of shaft openings, or manholes (fig. 76).[8]

An engineering variation on the principle of an underground canal with vertical shafts set at regular intervals can be found along sections of several canals that are constructed with open trenches up to 3.5 m long or even as deep open canals; such an unroofed section is called a *baṭṭ*. It seems that in these cases, the natural gypsum karst was too friable to allow proper tunnelling: the question of *shaṭṭ* or *baṭṭ* was decided based on the local geology. If a *baṭṭ* was built, bridging arches were left in place wherever feasible to provide stability and prevent wall collapse (fig. 77). It is also quite possible that the unroofed sections represent naturally eroded fissures in the karst that the engineers incorporated. This is suggested by the meandering course taken by some of the deep open canal sections that seem to follow an erosion pattern rather than a

7 Our local informant ʿAwaḍ Bā Ḥamīsh reported that during his time as a digger of underground canals, up to a hundred two-man teams were involved in the operation. One man dug the shaft and connected the gallery underground, while the other hauled up the spoil at the top. Two-man teams are only possible when the shafts are closely spaced; otherwise, a third worker is required to drag the full bucket along the gallery to the nearest shaft.

8 For the apt term "manhole," see Beadnell 1909, 177. Beadnell applies it in his description of subterranean canals in the Kharga Oasis in Egypt.

FIGURE 76 Row of rock-cut shaft openings north of Ghayl Bā Wazīr (50 cm scale)

properly surveyed line. The latter would result not in such meanders but in the perfect straightness that is clearly visible in most of the canal system.

Wherever a canal had to be dug through loose stones or rock, a potentially dangerous situation arose because of the risk of collapse. ʿAwaḍ Bā Ḥamīsh, the only digger of underground canals still alive in the Ghayl Bā Wazīr area, whom we had the chance to interview during our January 2001 field season, stressed that cutting the canal through rock was the preferred technique; digging through loose stones was only done if circumstances made it inevitable. In that case, the sides of the shafts and the canal had to be lined with locally made cement for stabilization (fig. 78).

FIGURE 77 Bridging arch along a deep open *maʿyān* section

After running below ground for, on average, several kilometres, depending on the terrain, the canal surfaces at its destination—that is, the fields and gardens irrigated by it—and continues as an open canal. The local population uses the term *maʿyān*, pl. *maʿāyīn*,[9] to refer to such a canal, which consists of an underground section (the *shaṭṭ*) and an above-ground section. Each one of the *maʿāyīn* has a name. For instance, the 6 km long Maʿyān al-Furāt ("Euphrates," fig. 79) has a 2.5 km underground section before it becomes a surface canal; as indicated by its name, it had a strong and reliable flow in the past. Maʿyān al-Ḥarth—fed by Ḥawmat al-Ḥarth—is traditionally considered

9 Serjeant (1964, 58) vocalizes the term *miʿyān*.

FIGURE 78 Cement reinforcement of a *maʿyān* shaft. Wind erosion has left the reinforcing collar around the shaft opening slightly elevated above ground level

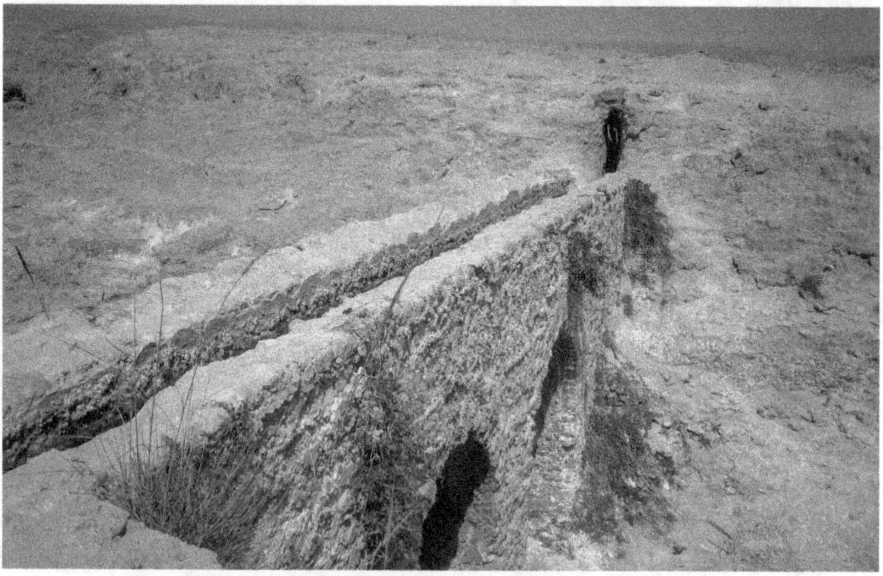

FIGURE 79 Maʿyān al-Furāt, with bridge across Wādī Shaqīb

FIGURE 80 Stone barrier (*mardaʿ*) with one opening (30 cm scale)

the oldest and largest in the area. The term *maʿyān* refers to both the canal itself and the land that it supplies with water, a linguistic detail that emphasizes their interdependence.

In order to regulate the flow rate of the canals, a stone barrier (*mardaʿ*, pl. *marādiʿ*) was often left intact in the last third of the underground section of a *maʿyān* before it comes to ground surface. This meant that the *mardaʿ* could be reached quite easily by the farmers. It typically had one to three round openings (*ḥarrāt*, sg. *ḥarra*) set into it, each with a diameter of 15–20 cm and placed one above the other, some 50–60 cm apart from centre to centre (fig. 80). By blocking and unblocking them individually, the farmers could use the *mardaʿ* as a sluice gate: during a heavy rain, they would close two or even all three holes, which reduced the water in the *maʿyān* to the overflow over the sluice gate and protected the irrigated gardens and fields from water damage. When the water level in the *ḥawma* was low, all three holes were opened to maximize the flow.

3 Other Water Sources

As our field study revealed, the groundwater in the collapse sinks may have provided the most obvious source of water that was tapped in Ghayl Bā Wazīr, but it was not the only one. For instance, Maʿyān al-Furāt was started with the intention of connecting it to Ḥawmat al-Kabsh. The vertical shafts had already been cut all the way back to the collapse sink when a strong underground source was accidentally struck some 500 m downstream from the *ḥawma*. Therefore, the gallery connecting the bottoms of the shafts remained unfinished beyond this point. Maʿyān al-Furāt is only one example that shows how people learned to exploit other water sources by taking a variety of approaches. However, the common element is the delivery of water to its destination by way of a *maʿyān*.

3.1 *Connecting a* Maʿyān *to a "Father Well"*

One principle that we documented in those parts of the study area where collapse sinks do not exist was the digging of a well-like trial shaft in search of groundwater. If the search was successful, the shaft was deepened and widened to allow maximum exploitation of the water and became what in Ghayl Bā Wazīr is called a "father well" (*ab*) that would feed a new canal.[10] For this, the same engineering steps were applied that have been described above—namely, digging vertical shafts at regular intervals and connecting them through a gently sloping gallery at the bottom, the *shaṭṭ* section of the *maʿyān*, to tap the well and convey the water to its destination under gravity flow.

The site of the trial shaft was chosen neither randomly nor by a water diviner. People took local surface clues as indicators of possible underground water resources. For instance, they paid close attention to water vapour (referred to locally as *kahra*) or condensation around the natural joints and fissures in the gypsum rock, particularly in the early morning when the air is moderately cool. As pointed out above, seepage is common in karst landforms, and firm evidence can be found in large gypsum crystals that form through the evaporation of water moving slowly through bedrock joints (fig. 81). They indicate

10 The term that is used in other parts of the Middle East is "mother well" (*umm*). See below, section 10.

FIGURE 81 Gypsum crystals that formed around a natural fissure (20 cm scale)

underground water not too far from the surface. In the local dialect, these natural joints and fissures that percolate water are called *khuwaʿ*, sg. *khawʿa*. People also observed variations in the natural vegetation. The fact that a shrub can seemingly grow out of the bare rock strongly suggests that water must exist not too far from the surface and can be reached by the plant's root system (fig. 82).

Of course, detecting groundwater at the bottom of a test shaft does not guarantee that the source will provide a dependable flow. Our informants explained that when water was reached, it would be bailed out quickly at intervals over a three-day period. If the flow proved unreliable, it was called *māʾ kādhiba*

FIGURE 82 Shrub growing out of a "father well" north of Ghayl Bā Wazīr

(literally, "lying water" or "false water"). In such cases, digging trial shafts would continue until one of them produced a sufficiently abundant water flow. We documented a site in Wādī al-Fijra, north of Ghayl Bā Wazīr (see fig. 68), where it appears that four shafts were dug before the engineer was satisfied. The possibility exists, though, that several shafts were completed and then connected underground to increase the flow in the *maʿyān*.

3.2 *Tapping Groundwater in the Wadi Bed*

Well-like test shafts could also be dug in a wadi bed. Here the water table is usually quite high because the groundwater is directly replenished by infiltration during the seasonal spate. If the probe located a satisfactory water resource, the shaft needed to be ringed with masonry to protect it from the flood and to keep out silt, coarse gravel, and boulders carried by the *sayl* (fig. 83).

3.3 *Diverting the* Sayl

A different way of making good use of the wadi and the seasonal spate was to divert part of the *sayl* directly into an underground canal. This required sinking a shaft to the side of the wadi bed, filling it with stones, and constructing a "fenestrated box" around its opening, through which the water was forced to pass in order to filter out as many particles carried by the *sayl* as possible. In this way, only relatively pure water entered the *maʿyān* system;

FIGURE 83 Rock-cut canal at the wadi edge, with masonry-ringed probe shaft (*top right*)

otherwise, the underground canals would have become clogged sooner rather than later.

One of the major canals, Maʿyān Bā Mardūf, is an example of this principle. Its source is a father well 30 m from the southern bank of a small wadi, the Wādī Shaqīb. The upper course of Maʿyān Bā Mardūf runs parallel to the wadi for some 400 m, and it is at the end of this stretch that a shaft with a fenestrated box was dug right at the edge of the wadi. During a *sayl*, it supplied Maʿyān Bā Mardūf with additional water, thereby leading to a seasonal increase of the flow.

3.4 Maʿyān *Junctions*

As a result of the various approaches taken to exploit the available water resources, the landscape of Ghayl Bā Wazīr is crisscrossed by a dense network of canals. In particular, the western part of the study area evokes a complex railway junction (fig. 84). At four-way junctions, where one *maʿyān* had to cross another, the engineers constructed the upper canal's floor as a bridge over the lower one (fig. 85).

4 *Maʿyān* Gradients

Whatever the source of water that was tapped, a proper gradient was essential for the functioning of the *maʿyān*. Water flowing by gravity can overcome occasional unevenness of the canal floor so long as the current is sufficiently strong. A gradient that is too steep increases the risk of erosion damage to the canal; it also leads to overuse of the water resources to the point where the system may eventually run dry because water is drawn at a rate that exceeds the natural recharge, a planning error with far-reaching consequences for the entire community. From our fieldwork, it became obvious that, overall, the engineers in Ghayl Bā Wazīr worked with precision.

Unfortunately, there are few figures to cite. In January 2001 when we started our first field season, major parts of the *maʿyān* system had recently fallen into disuse because overexploitation of the aquifer by deep wells had caused the water table to drop dramatically. A number of the collapse sinks and the canals connected to them had run dry, resulting in very serious problems for the local population (see section 12). Debris accumulated in the derelict canals made it impossible for us to record precise measurements except where diesel pumps had been installed to tap the groundwater from its current, much lower, level and feed it back into the *maʿyān* system. This practice could be observed in the middle part of Maʿyān al-Furāt and allowed us to measure its gradient as 0.03% (a drop of 0.47 m in 1.5 km). Clearly, *maʿyān* engineering involved accurate surveying and execution of the work.

5 Water Use

Studying the engineering principles and mapping the canals also gave us the opportunity to observe the different uses of the water. Along the lower course of a *maʿyān*, we find the typical structures that made it possible for people to exploit the water. Stairs leading down to the canal were built at intervals

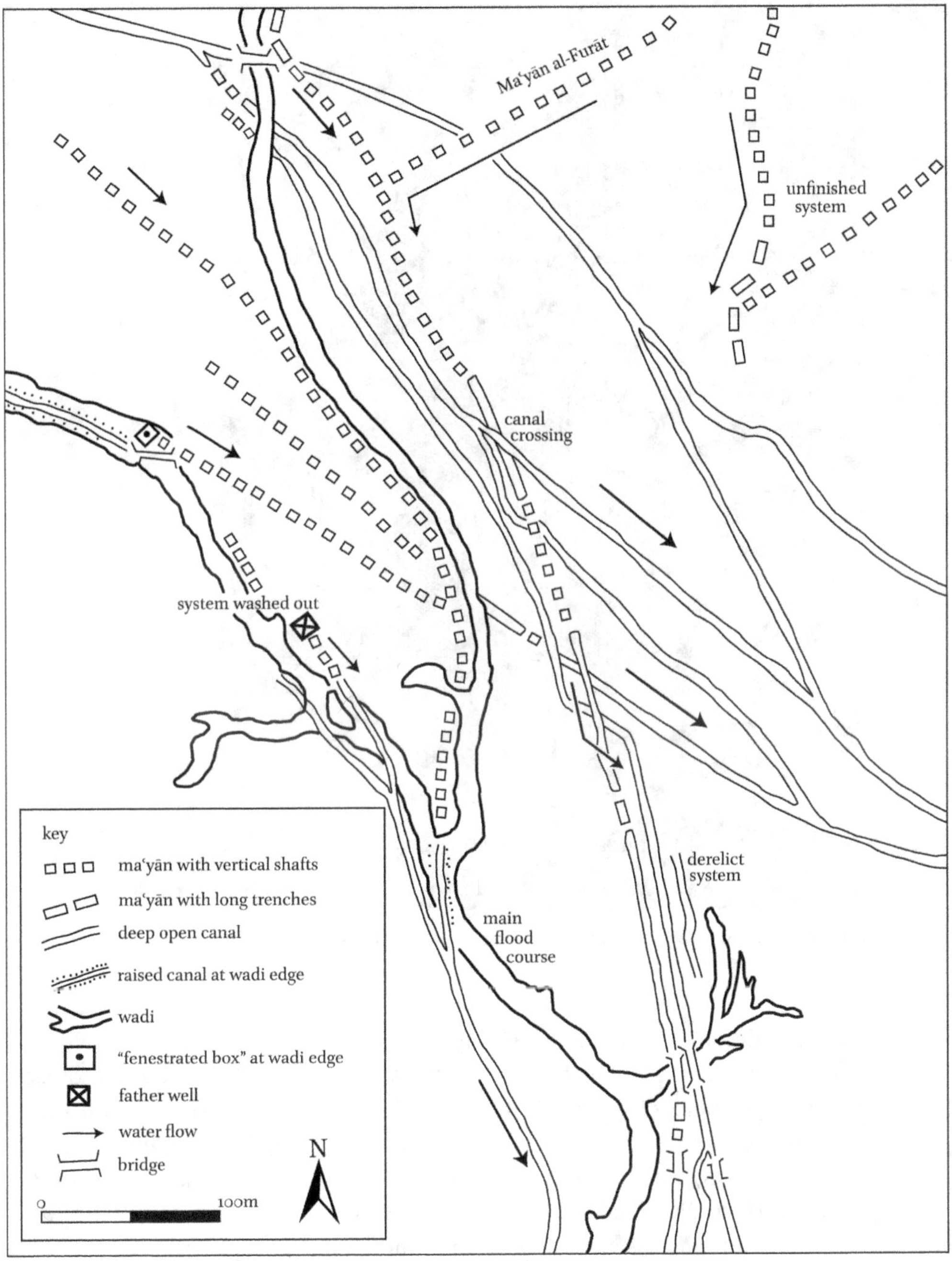

FIGURE 84 The busiest part of the *maʿyān* crossings in the central-western area of the study region

FIGURE 85 Four-way junction, with one *maʿyān* crossing another as a bridge

and provided free access for fetching water (fig. 86). Public spaces for doing laundry, gender-segregated bathing places, ablution facilities next to mosques, and watering places for animals were supplied by channelling off water from a *maʿyān*. Some of the laundry, bathing, and ablution facilities continue to operate today (fig. 87). It is important to note that the water is considered communal property; individual houses do not have private access to it. After a *maʿyān* has reached the fields and gardens and emerges at ground surface, the open canal bifurcates repeatedly to irrigate them.

It was the location of the cultivable land that determined the course—and to a certain extent the gradient—of a *maʿyān*. The settlement developed close

FIGURE 86 Stairs leading to a *maʿyān* in the town of Ghayl Bā Wazīr

to the land along the lower course of the underground section of the canal—that is, towards the southern end of the gypsum shield. Accordingly, the water was available for domestic purposes when it was still clean, and it was easily accessible since it was flowing at a rather shallow depth. When the *maʿyān* had traversed the town below ground, its water was no longer suitable for domestic use but was perfectly adequate for irrigating the vegetable gardens and orchards beyond the town (fig. 88). Allocating the water first as a clean resource to the settlement and then as grey water to the fields is effective use of the precious commodity. Today, as in the past, fruit such as dates and pomegranates, vegetables, salad greens, and fresh fodder for animals are cultivated in well-kept market gardens; in addition, tobacco and henna are grown as cash crops.

FIGURE 87 Ablution facilities of a mosque, supplied by a *maʿyān* branch

Ghayl Bā Wazīr is known for the good quality of these last, which thrive in the calcium-rich soil. In places where the soil was judged to be too poor, topsoil was brought in by donkey or camel, often from several kilometres away.[11]

Before the installation of a modern water-supply network in the 1970s, the inhabitants of Ghayl Bā Wazīr preferred not to drink from the underground

---

11 One also needs to remember that the *maʿyān*—unlike the *sayl*—carries hardly any suspended matter and therefore does not contribute to maintaining or improving soil fertility.

FIGURE 88 Irrigated market garden

canals because the water flowing from the karst contains a high level of calcium. The exception was a *maʿyān* drawing sweet water from gravel beds to the west of the gypsum shield. Access to it for collecting drinking water was restricted to prescribed hours of the night, a regulation thought to minimize contamination by animals and humans. After all, keeping the drinking water clean was of utmost importance.

6 Timing Water Allocation

A *maʿyān* was usually owned by a group of people on a time-share system. Since the flow of water was continuous, an issue of central importance was the way in which the scarce resource was allocated equitably and reliably to the different users. How were day and night divided into units of time before mechanical clocks were introduced to the area in the 1960s?

Traditionally in the Middle East, the day is not partitioned into hours of equal length as measured by a mechanical clock, but rather the daylight period is divided astronomically (i.e., based on the sun's daily passage through the sky) into twelve hours starting at sunrise and ending at sunset. Due to the fact that at any given locality the length of the daylight period changes over the course of the year, the length of the daylight hours varies as well: during the winter, one hour is shorter than during the summer.[12] Therefore, astronomical division of the day results in "seasonal" or "unequal" hours, with a duration of one twelfth of the daylight period.[13]

The length of such an hour was determined by using the sundial principle—that is, by observing the shadow cast by an object placed in the sunshine. In Ghayl Bā Wazīr, a farmer's shadow was measured at a given time of day, and a prescribed increase (or decrease) indicated a unit of time. The shadow lengths at a location vary over the course of the year because of the sun's changing altitude. In Ghayl Bā Wazīr, therefore, shadow lengths were determined at specified times of day throughout the year and recorded in a chart for every sixth or seventh day of the year.[14]

During the night, water allocation was timed with the help of a star calendar. The underlying principle was observation of the moon's path against the background of the fixed stars. Each successive night in the course of a lunar month, the moon shifts to a new position in relation to the background stars. Twenty-eight conspicuous stars or constellations, known as the "lunar mansions," *manzil* (or *manzilat*) *al-qamar*, pl. *manāzil*, were identified along the moon's path, and thirteen or fourteen of them are visible during any given night of the year, in accordance with the season. It is the succession of these lunar mansions that was used to measure time during the night. Depending on the size of his land, a farmer had one or more designated mansions, and his turn for irrigation started as soon as that constellation had reached its highest position in the sky, the zenith. His time was up when the lunar mansion belonging to the next farmer in the irrigation cycle had succeeded to that position.[15] He then blocked his canal inlet with a small temporary dike called a *ṣaqīʿ*, pl. *ṣuqūʿ*, which typically consisted of stones, mud, grass sods, and cloth.

---

12 The difference in duration of such a daylight hour depends on the latitude of the locality; at more northerly (or southerly) latitudes the seasonal variation is much greater than close to the equator.

13 King 1990, 250; see also King 1996, 172.

14 The astronomical details are described in Hehmeyer 2005, 87–96.

15 See ibid., 89–92, for further details. The principle based on lunar mansions is not nearly as well known as the one that observes the sun's annual path against the background of the fixed stars and divides it into twelve equal zones, each covering thirty degrees and marked by—and named after—a conspicuous constellation serving as a zodiac sign, e.g., Gemini or Cancer. This alternative concept originates in ancient Mesopotamia and was adopted and refined by the Greek astronomers, who passed it on to the Western world. The popularity of the solar zodiac is surprising insofar as it requires the ability of abstraction, since the stars are invisible when the sun is in the sky. This is not the case with the moon's path against the background of the fixed stars: it is easy to observe because the constellations are visible together with the moon.

For reasons of fairness, the irrigation schedule for a specific *maʿyān* had an odd number of divisions in order that a farmer's time slot would rotate between day and night. Having received water during the day in the previous irrigation cycle, the farmer's next time slot would be during the night. This regulation takes account of the differences in length between daylight and night hours and ensures an equitable water share for each farmer over the course of the year.[16] Exceptions to the rule were those *maʿāyīn* that had a rather weak flow, too weak to be used directly for irrigation. In these instances, the water was collected during the night in a reservoir (*jābiya*, pl. *jawābī*). Its outlet, locally called *ʿayn*, was plugged at the time of the *maghrib* prayer—that is, at sunset—by a *khudra*, which is a piece of cloth attached to a rope that could be pulled the next morning at the time of the *fajr* prayer—that is, at daybreak—to release the water.

If the ratio of unit of time to unit of land is the same all along the canal, measuring a water share in time instead of volume can result in considerable inequalities depending on the exact location of the plots. Close to the spot where the *maʿyān* flows out into an open surface canal, the flow rate is much stronger than at the lower end; water is lost in transport through evaporation and, more importantly, through seepage along the course of the earth canal. To compensate for the disparity, in the traditional system, farmers used to own land in different parts of a *maʿyān*: at the high end, in the middle region, and at the lower end.[17]

Timing water allocation was the responsibility of the *muqaddam al-maʿyān* (supervisor of a specific canal).[18] As a landowner and farmer in the area irrigated by that *maʿyān*, the *muqaddam* was elected by his fellow farmers not only for his astronomical knowledge and experience, and because he was trustworthy and responsible, but equally for his leadership qualities and good judgement. He also monitored regular maintenance work, to which each farmer contributed labour (and, from time to time, also cash) in proportion to his water share. As well, the *muqaddam* had the authority to remove a field that had been neglected by its owner from the irrigation cycle; not cultivating one's land could lead to the loss of entitlement to water. As compensation for his work, the *muqaddam* received an additional amount of water. Depending

16 While one might consider irrigating during the night as inconvenient, the night is, in fact, the preferred time for irrigation: under the climatic conditions that prevail in the Ghayl Bā Wazīr area, daytime temperatures are very high.

17 Differences in flow rate are also noticeable from one *maʿyān* to another, which is why the ratio of unit of time to unit of land differs between them.

18 The term was erroneously vocalized *muqaddim* in a previous publication (Hehmeyer 2005, 89).

on the size of the *maʿyān*, the *muqaddam* had three or more assistants, who formed an advisory committee (*lajna*). For instance, Maʿyān al-Ḥarth, which is considered the largest of the *maʿāyīn* both in terms of the flow of water and the area irrigated by it, was managed by the *muqaddam* with the help of nine assistants working in three shifts. Interestingly, the Arabic root *q-d-m*, from which the word *muqaddam* is derived, has ancient origins, and an Ancient South Arabian derivate with the meaning of "canal supervisor" can be found in several inscriptions.[19] See Case Study 1, section 6.1, for the stresses of being the canal supervisor in ancient Mārib.

During our field research in December 2003–January 2004, we were fortunate to meet a former *maʿyān* supervisor, Muqaddam Sālim, who was from the Ghayl Bā Wazīr area and was willing to share his expertise with us. He stated with confidence that on days when the sky was cloudy, he was able to judge time based on his experience. Muqaddam Sālim also described the potential vulnerability of the irrigation scheme and emphasized the significance of consensus among the farmers with regard to water allocation and the management of the *maʿyān* system. He pointed out that they had sufficient common sense to avoid manipulation for short-sighted advantage.

The earliest written record of the timekeeping practices is a manuscript that indicates a sixteenth-century origin for the measuring method.[20] The historical sources do not give any evidence for disputes over water shares in Ghayl Bā Wazīr, and the fact that regulations from the sixteenth century provided the framework for water allocation until the 1960s underscores that the community accepted them as being fair. Interestingly, an astronomical almanac based on lunar mansions is published annually by the local authorities to this day. Its main purpose is the determination of prayer time. In this way, at least part of the tradition is kept alive.

7 The Origins of the *Maʿyān* System

It is Shaykh ʿAbd al-Raḥīm ibn ʿUmar Bā Wazīr who is credited with having founded the town of Ghayl Bā Wazīr shortly after his arrival in the region in 1307/8.[21] The entrance to the congregational mosque of Ghayl Bā Wazīr bears an inscription stating that it was established by him in 1317/18 (fig. 89). According to one local legend, he discovered the subterranean water resources

19 Stein 2010b, 337–344.

20 See Hehmeyer 2005, 92–93, for details.

21 Ibn Shaykhān 1999, 19 and 22.

FIGURE 89 Entrance to the congregational mosque in Ghayl Bā Wazīr with foundation inscription

by firing an arrow into the air: it landed on a spot from which a spring immediately flowed forth. Of course, stories reporting the miraculous detection of water are common in arid lands. This version, though, reflects the geological reality of the terrain and the fact that in the karst landform of Ghayl Bā Wazīr, the sudden appearance of water from below ground was—and still is—nothing unusual. ʿAbd al-Raḥīm ibn ʿUmar Bā Wazīr is also recognized as having dug the first *maʿyān*, Maʿyān al-Shaykh, to supply water to the mosque, which underlines the need for water in an Islamic town to fulfil the religious requirement of proper ablution before prayer. It is worth noting that the site of the mosque and the adjoining cemetery is located in what is now the lowest part of the town. Clearly, as the town developed and grew since the mosque's foundation, the ground surface rose over time as a result of the accumulating urban waste. Only the mosque and the cemetery remain more or less at the original level. When ʿAbd al-Raḥīm ibn ʿUmar Bā Wazīr died in 1346, he was buried in the mosque, where his tomb can be visited today.[22]

A sample that was taken from the sediment deposits at the bottom of one of the collapse sinks, Ḥawmat al-Ḥarth (see fig. 68), gives an (uncalibrated) radiocarbon date of 600 BP—that is, 600 radiocarbon years before the present. Calibration results in a calendar date falling within the fourteenth century, and this seems to be more than a coincidence. As described above, the underground water movement in karst landforms erodes the bedrock and, over time, leads to caverns filled with clean groundwater, filtered by its passage through the rock. Once the roof of such a cavern collapses, wind-borne material and aquatic organisms start to accumulate at the bottom of the now-open reservoir. These deposits are excellent archives of past events. The calibrated date tells us that Ḥawmat al-Ḥarth has been exposed since at least the fourteenth century. We may safely assume that in as arid an environment as the Ghayl Bā Wazīr area, people would have tried to find a way to exploit that water from its sudden appearance. Of course, one does not need to exclude the possibility that other collapse sinks are older and were tapped before the fourteenth century.[23]

According to a different local legend, it is Maʿyān al-Ḥarth, fed by Ḥawmat al-Ḥarth, that is considered the oldest—and largest—*maʿyān*. Its construction is attributed to a Jew, an interesting nuance insofar as credit for a massive engineering scheme is given to someone who is not a Muslim. Again, stories following this pattern are very common in premodern society. They attribute to "the other" powers of an unknown extent and an unknown, perhaps even magical, nature, which allow him—and less frequently her—to achieve what everybody else could only dream of.

---

22 For the interesting family background of Shaykh ʿAbd al-Raḥīm ibn ʿUmar Bā Wazīr, see Ibn Shaykhān 1999, 19–23. "Shaykh" does not imply tribal-chief nomenclature in this context but reflects the family's hereditary nobility; see Naval Intelligence Division 1946, 404.

23 Archaeological evidence for much earlier occupation of the region is presented in Hehmeyer, Keall, and Rahimi 2002, 92–94.

8 Problems with the Water Table

In the Ghayl Bā Wazīr area, the rising and falling of the water table can be directly observed in the water level in the collapse sinks and in the flow rates of the canals. Unusually heavy rainfall in the catchment area leads to an increase in the underground flow of water, to the point where this may cause damage to the *maʿyān* system. The lowering of the water table, on the other hand, can have a more dramatic effect, in particular as soon as the point is reached where a *maʿyān* can no longer be fed from a *ḥawma*.

Ḥawmat al-Ḥarth serves as an example. Two *maʿyān* openings are visible in its southern face (fig. 90). The eastern one (*A*) is the original outlet of Maʿyān al-Ḥarth, constructed at this point as a deep open canal, with its base at 5.1 m below ground surface. A second outlet to the west (*B*) has its base a further 4.36 m below the original one and forms the entrance to a gallery (ca. 0.9 m high) that connects with Maʿyān al-Ḥarth some 70 m along its course. Clearly, construction of the new branch became necessary to feed water back into the system because the water level in the *ḥawma* had dropped and the original canal could no longer be reached. The local term for a feeder canal that connects an additional water source (or reconnects the original water source) to an existing *maʿyān* suffering from reduced flow is *taqdūma*.

Interestingly, it appears as though the engineers of Maʿyān al-Ḥarth had anticipated a falling water table by designing a canal that would accommodate a lower intake at some point in the future. For the first 43 m of its course, the original *maʿyān* is extremely uneven. Its bottom rises and falls until it is almost 0.5 m above the outlet at the *ḥawma*. After that, it drops sharply more than 5 m, yet this point is more than 25 m before the *taqdūma* connects to the *maʿyān* 70 m downstream. We also observed such stepped bottoms in several other underground canals.

It is not clear what caused the problem with the water table, and whether a prolonged drought or overexploitation of the water resources was to blame. Quite clearly, in some of the collapse sinks, the groundwater was not replenished at the same rate as it was drawn. Two of the smaller collapse sinks, Ḥawmat al-ʿArus and Ḥawmat al-Kabsh, dried up completely many years ago, and none of our local informants remembered having seen water in them. However, according to a written report, Ḥawmat al-ʿArus contained water in the year 1920, with its level at 10 m below ground surface.[24]

24 Little 1925, 55. The author does not mention the collapse sinks (for which he erroneously uses the term *miʿyān*) by name, but his description of three that he visited clearly refers to Ḥawmat al-Ḥarth, Ḥawmat al-Sirkāl, and Ḥawmat al-ʿArūs. Little reports the same water

FIGURE 90 Ḥawmat al-Ḥarth: southern face with two *maʿyān* openings, one to the east (*A*) and one to the west (*B*)

It is obvious, then, that a falling water table was not an entirely unfamiliar problem in the past. As a consequence, the water flow in a *maʿyān* could decrease to the point where it might dry up completely because the water in the *ḥawma* no longer reached the *maʿyān*. In that case, a canal had to be deepened along its entire length in order to maintain a continuous flow, a practice referred to locally as *sawqa*. Alternatively, a new feeder, the aforementioned *taqdūma*, was connected to the original *maʿyān*. Local storytelling is full of examples of such emergency action. Even though the challenge was enormous, it was manageable—a surmountable obstacle for people who were experienced in planning and digging underground canals. The agricultural necessity for water meant that the problem was tackled by community effort, based on social consensus. Nowhere in the historical documents is there any evidence for serious controversy, and our local informants agreed that water-management issues were resolved by mutual agreement. Without water, there was no way for the community to survive.

---

level of 10 m below ground surface in all three of them, which is considerably higher than the water level in Ḥawmat al-Ḥarth and Ḥawmat al-Sirkāl at the time of our first field survey in January 2001. We measured an overall depth of the dry Ḥawmat al-ʿArūs of some 14 m.

9 Reports on Ghayl Bā Wazīr in Western Sources

The earliest first-hand account of Ghayl Bā Wazīr in Western literature is by Leo Hirsch, based on his visit in August 1893 and published in 1897.[25] We owe the first brief description of the underground water system to Theodore Bent, who came to Ghayl Bā Wazīr in February 1894 with his wife, Maybel.[26] In the travelogue published by Maybel in 1900, after her husband's death, nothing substantial is added to the original record of the site except a reference to the legendary discoverer of the underground water, whom she names "Shaykh ʿUmar."[27] Apart from Little's 1925 report on the geology of the area, cited at the beginning of this chapter, the next significant contribution occurred in a rather convoluted way. At a Royal Geographical Society meeting in 1930, W. H. Lee Warner made some after-lecture comments in response to Squadron-Leader R. A. Cochrane's showing an RAF aerial photograph of the town of Ghayl Bā Wazīr, which had been taken in 1929 during an air reconnaissance of the Ḥaḍramawt. Lee Warner's postlecture remarks were published in 1931, including a reference to the photograph and some details concerning the collapse sinks and the underground canals.[28] As later examples of brief notes following short visits, we may cite the first-hand observations of the Ghayl Bā Wazīr water systems by Robert B. Serjeant[29] and, most recently, Dale Lightfoot.[30]

10 The Question of the Iranian Connection

Invariably, visitors observe the resemblance between the *maʿyān* system of Ghayl Bā Wazīr and the well-known Iranian underground water systems called

25 Hirsch 1897, 277–280. The even earlier short reference to Ghayl Bā Wazīr by van den Berg (1886, 80) is contained in his work on the Ḥaḍramawt that he compiled from information provided by Hadrami emigrants in the Dutch East Indies. Van den Berg was resident in Batavia and never visited the Ḥaḍramawt himself.

26 Bent 1894, 317.

27 Bent and Bent 1900, 200.

28 Lee Warner 1931, 221. While Cochrane showed the photograph during the meeting in 1930, he did not reproduce the image in his own published text (Cochrane 1931, 209–216). The image was published by Daniel van der Meulen (1932, 391), who visited Ghayl Bā Wazīr in June 1931 with Hermann von Wissmann; see van der Meulen and von Wissmann 1932, 26–28. However, the text that accompanies the photograph is copied verbatim from Lee Warner's original description.

29 Serjeant 1964, 57–58.

30 Lightfoot 2000a, 14–15.

qanawāt, sg. *qanāt*.[31] A *qanāt* is designed to tap underground water generated by precipitation falling on high ground adjacent to arid plains where agriculture depends on irrigation. It is typically dug through an alluvial-fan outwash that forms at the foot of the mountain range. The alluvial sediments have a high infiltration capacity. Where they overlie impervious bedrock, they constitute an excellent water-bearing stratum, a more or less shallow aquifer that is naturally recharged by rain or melting snow from the mountains. The first step in *qanāt* construction involves digging well-like test shafts in order to find out whether the groundwater resources justify further investment. If the probe is successful, a large-sized shaft is sunk into the aquifer; it becomes the "mother well" (*umm*) of the new system.[32] Following a meticulous survey of the terrain from the mother well to the point of destination, an almost horizontal canal is tunnelled back through the alluvial fan to tap the mother well and lead the water under gravity flow to its destination. The gradient should be very slight, between 1:1000 and 1:2500, to prevent erosion—and possibly the collapse—of the largely unlined walls. Since the workers need access, vertical shafts are dug at intervals, on average 18–135 m apart. The shafts also facilitate removal of the spoil, which is dumped around the shaft openings, creating the line of "manhole spoil heaps" that is so typical of a *qanāt* and allows for easy identification, especially from the air. Eventually, the *qanāt* comes to the surface and continues as an open canal.

Quite obviously, with regard to the engineering principles involved, *maʿyān* and *qanāt* show strong similarities. It is therefore not surprising that Bent refers to the underground canals of Ghayl Bā Wazīr as being "similar to the Persian kanats,"[33] but he does not state that they actually are *qanawāt*. However, others have taken these observations and read more into them than is actually there. Henri Goblot can be taken as a fine example: "En 1899 [*sic*], le voyageur Th. Bent a vu des qanats à Ghaïl Balwazir [*sic*]."[34]

31 The following brief description is based on Lambton 1978, 529a–531a; Ron 1989, 231–233; and Wilkinson 1977, 76–81.

32 Lambton (1978, 529b–530a), Ron (1989, 232–233), and Wilkinson (1977, 76–80) explain how the location of the mother well is determined through a compromise between establishing a stable groundwater source, keeping the costs for digging as reasonable as possible, and satisfying the water requirements at the point of destination.

33 Bent 1894, 317.

34 Goblot 1979, 105. Unfortunately, this quotation is also an example of the numerous factual mistakes that are contained in Goblot's work. Typical of the trend to read one's own pet ideas into another person's text is Goblot's distortion of the 1871 report made by S. B. Miles and M. W. Munzinger on the Ḥaḍramawt, from which he extrapolates what he calls the first report of a *qanāt* (105), while the authors simply describe having "passed through a subterranean passage, about 150 yards long and 7 feet high, apparently excavated

It was the British archaeologist Gertrude Caton Thompson who took the issue one step further and first attributed the construction of the underground canals of Ghayl Bā Wāzir to the help of Iranians. In a 1944 publication, she surmises that the irrigation technology in southwest Arabia in general "arose and evolved under foreign inspiration," and that "Arabia's contribution to the world's material progress has presumably always been negligible."[35] Even though she fails to provide any factual evidence, she describes the underground water systems on the coastal plains of Yemen and farther inland as directly derived from the *qanāt* technology of Iran and suggests that they were introduced during the Achaemenid presence in ancient Arabia.[36] Building on this supposition, Dale Lightfoot, who visited Ghayl Bā Wazīr in the late 1990s, makes the bold statement that "Sassanid Persian engineers were designing *qanats* in the Hadramaut (probably using imported *moqani qanat* builders)."[37] Credit is no longer given to the Achaemenids but rather to the Sassanians who controlled South Arabia from around 575 to shortly after 628 CE. When considering the channelling of underground water, Westerners seem to have little confidence in the technological skills of anybody but Iranians.

Yet the Iranian *qanāt* and the southern Yemeni *maʿyān* are not identical. Most importantly, while the former is dug through alluvial sediments to tap the aquifer in the piedmont, the latter is cut through bedrock, typically back to an open collapse sink. According to the professional *maʿyān* digger ʿAwaḍ Bā Ḥamīsh from Ghayl Bā Wazīr, digging through alluvium was only done if circumstances made it inevitable; his Iranian counterparts would even change the course of a *qanāt* when they encountered rock in order to avoid it.[38] Engineering a *qanāt* through rock is not the standard Iranian approach.[39]

One also needs to remember that in a karst landform, seepage out of natural joints and fissures is nothing unusual; as described above, the local dialect even uses a specific terminology to describe the phenomenon. Digging back was done in search of underground water. Technically speaking, this would relate the *maʿāyīn*—in part—to what are called "spring-flow tunnels," the building

artificially for the purpose of bringing water from the nullah for storage." See Miles and Munzinger 1871, 221; *nullah* is the term that is used in South Asia for "wadi."

35 Caton Thompson 1944, 9–10.

36 Ibid., 10, n. 1.

37 Lightfoot 2000b, 222.

38 I had the chance to talk to four professionals, *qanāt* masters, in Yazd in November 2015 on the occasion of the *Second International Water Association Workshop on the Evolution of the Qanat and Related Hydraulic Technologies* at the International Centre on Qanats and Historic Hydraulic Structures (under the auspices of UNESCO). All four agreed that one should keep clear of rock, if at all possible.

39 See English 1968, 176: "Qanat construction in solid rock is rare."

of which was well established in Palestine in the first millennium BCE.[40] The digging back can be continued for a considerable distance and may therefore involve access and ventilation shafts.[41]

In hydraulic engineering, with regard to the technical approach taken, the decisive factors are the hydrogeological and geomorphological characteristics of the local environment.[42] Accordingly, construction details of underground canals vary throughout the Middle East and North Africa, and so does the terminology. What is called a *qanāt* in Iran or a *maʿyān* in southern Yemen is referred to, for instance, as a *falaj* in Oman, the United Arab Emirates, and Saudi Arabia; a *fujjāra/fuqqāra* (foggara) in North Africa (especially Libya and Algeria); a *kārīz* in eastern Iran and Afghanistan; and a *khaṭṭāra* in Morocco.[43] The literature uses the umbrella term *qanāt*.

To date, the earliest archaeological remains are from the southeastern Arabian Peninsula, in the territory of the modern-day United Arab Emirates. Examples from the Emirate of Abu Dhabi are well documented and confirm the *falaj* systems (as they are termed locally) date to the early first millennium BCE.[44] For instance, *falaj* system Hīlī 15 provided water for nearby sites and settlements that can be dated based on the pottery found during the archaeological work; it stems exclusively from the early first millennium BCE.[45] According to archaeologist Rémy Boucharlat, who surveyed one of the sites, *falaj* Hīlī 15 "belongs without any doubt to this period."[46] Another example, a *falaj* system at Bidaʿ Bint Saʿūd, is dated once again based on pottery from the settlement that it supplied with water; in addition, several hundred pottery sherds were excavated from the bottom of a stairway providing access to the gallery. As

40 Ron 1989, 219–234. The general principle has also been described by Beaumont (1989, 17).

41 Ron 1989, 225 and 233. See Semsar Yazdi and Labbaf Khaneiki 2010, 62, for an example from modern-day Iran.

42 Costa 1983, 275; and Ron 1989, 231–232.

43 See Beaumont, Bonine, and McLachlan 1989, x–xv.

44 Al Tikriti 2002, 117–138; and Al Tikriti 2011, 67–135.

45 Al Tikriti 2011, 67–83.

46 Boucharlat 2003, 166. The surveyed site in question is Hīlī 14. Boucharlat's hypothetical distinction between a *falaj* or *qanāt* proper that taps a deep aquifer on the one hand, and what he refers to as a "water-draining gallery" that taps shallow water on the other (162–163), seems artificial and confusing, and as Boucharlat himself points out, it at least partly "rests upon a lack of information" (163). His categorization of Hīlī 15 (and Bidaʿ Bint Saʿūd, see below) as a water-draining gallery (166–167) should be understood as reflecting what he calls "a preliminary picture" that "will certainly change" once the publication of the excavation data is complete (164). That publication has since been completed; see Al Tikriti 2011. Based on the engineering principles involved, Hīlī 15 is a *falaj* or *qanāt*. Local variations are the result of specific environments, as I have outlined in this chapter. See also Al Tikriti 2011, 130–132, for a refutation of Boucharlat's hypothesis.

Walid Yasin Al Tikriti points out, this is "an ideal place to recover dating evidence since it is a communal area used for collecting water, usually by women or their young daughters, who may accidentally break a jar, leaving precious evidence behind for archaeologists to find."[47] Other examples exist in the Emirate of Abu Dhabi, although they are not as well documented.[48] However, at least one further site from the Emirate of Sharjah corroborates the dating of the *falaj* systems on the Arabian Peninsula to the early first millennium BCE.[49]

As for written evidence, the earliest reference can be found in a report by the Greek historian Polybius on the military campaign of the Seleucid king Antiochus III against the Parthians. It attests to authentic *qanawāt* on Iranian territory towards the end of the third century BCE, with the implication that they had been built several generations ago. Polybius gives a physical description of subterranean canals in the Iranian desert, including the well-like shafts that the Parthians filled in to deprive the advancing Seleucid troops of access to water; they then destroyed the outlet openings.[50]

There is no textual evidence of *qanawāt* for earlier times, even though the cuneiform text recording the eighth campaign of the Assyrian king Sargon II against the Urartians in the year 714 BCE has been repeatedly cited. The text is dedicated by Sargon to the god Aššur in celebration of his military victory. The translation by François Thureau-Dangin was published in 1912, and from lines 199–232 we learn of Sargon's having destroyed the palace and looted the storehouses at a site by the name of Ulḫu and of his boast that he had laid waste to Ulḫu's fertile land. The text informs us that Ulḫu's prosperity was due to an elaborate irrigation system. There is, however, nothing at all presented by Thureau-Dangin that identifies the canals as *qanawāt*.[51] Subsequently, Jørgen Laessøe tentatively interpreted the damaged lines 202–204 of the text as implying that the irrigation system at Ulḫu used *qanāt* technology,[52] but he called for future archaeological and textual investigation to verify what he himself described as a "theory."[53] He also identified Ulḫu as being located in what would be modern-day northwestern Iran.[54]

47 Al Tikriti 2011, 97; for the archaeological context, see 86–105. See also Boucharlat 2003, 167.
48 Al Tikriti 2011, 106–111.
49 This is the site of al-Madam (Thuqayba), which was excavated by the Spanish Archaeological Mission; see Córdoba 2013, 142–144 and 147–150. Al Tikriti (2011, 112–114) provides further archaeological details.
50 Polybius/Paton, trans. 1922–1927, 4.10.28.
51 Thureau-Dangin 1912, 32–39.
52 Laessøe 1951, 22–27.
53 Ibid., 22 and 31.
54 Ibid., 21.

Unfortunately, Laessøe's theory was wholeheartedly adopted as fact by Robert J. Forbes in his groundbreaking work *Studies in Ancient Technology* (1955).[55] Since then, Sargon's eighth campaign has been persistently invoked as providing evidence for the origin of the *qanāt*—for instance, by the authoritative Ann Lambton, who refers back to Forbes and makes the straightforward statement that the *qanāt* originates in northwestern Iran.[56] The aforementioned Henri Goblot confirms that "without doubt, Sargon saw *qanawāt*," citing Laessøe as his main source,[57] and because the water system was fully developed in the eighth century BCE, he suggests that its origins should be sought even earlier. In his 1990 book, *The Arabian Gulf in Antiquity*, Daniel Potts refers to Goblot for "the earliest generally admitted date for the beginnings of *qanat* irrigation in Iran, Armenia and Assyria, which is normally put between the tenth and eighth centuries BC."[58] Citing Forbes, Goblot, and Potts, Lightfoot makes a clear statement on the sole Iranian origin of the *qanāt* "some time between the tenth and the eighth centuries BC."[59] Many others have also uncritically used the same references in support of a *de rigeur* ancient Iranian origin of the *qanāt*.[60] The inclination to automatically credit the invention of this technology to the engineers of pre-Islamic Iran, without convincing evidence, has been aptly referred to by the archaeologist Paolo Costa as the "Iranian syndrome."[61]

Yet despite a century of scholarship, the toponyms of Sargon's eighth campaign are still much under debate,[62] which means that the location of Ulḫu cannot be identified conclusively—and therefore be cited in support of an ancient Iranian origin of the *qanāt*. In addition, the question of geographical origin is intertwined with the complex political history of the area and does not allow a straightforward ethnic attribution. Even though a site may be located in modern-day Iran, this does not mean that whatever originated there in the eighth century BCE was necessarily invented by Iranians. One should also not forget that, as Boucharlat points out, "there is no archaeological evidence of *qanāt* or any kind of underground galleries on the Iranian plateau in

55 Forbes 1955, 152–153.
56 Lambton 1978, 529a; and Lambton 1989, 5.
57 Goblot 1979, 67–68 and n. 6.
58 Potts 1990, 392.
59 Lightfoot 2000b, 217–218.
60 See Al Tikriti 2011, 136–143, for examples and further discussion.
61 Costa 1983, 275.
62 See, for instance, Vera Chamaza 1994, 91–118, and Vera Chamaza 1995–1996, 235–267, for an extensive bibliographic survey. Vera Chamaza suggests that Ulḫu was northwest of Lake Urmia (1995–1996, 237–238 and fig. 1), while Muscarella (1986, 468–471 and fig. 3) proposes a location southwest of it. Bagg (2000, 130–132) contributes further details.

antiquity."[63] But most importantly, a new analysis of the cuneiform text by Ariel Bagg has provided ample evidence—both linguistic and technological—that an identification of the canals with *qanawāt* cannot be supported; Laessøe's tentative interpretation of the damaged cuneiform text was incorrect.[64]

11 Conclusion

The *maʿyān* of southern Yemen—like the *qanāt* of Iran—is a technical response to specific hydrogeological and geomorphological conditions. The *maʿyān* suits a particular environment, as do the other local variations listed above. Its great benefit is that it provides a dependable supply of water year-round. The value of the water is explained by both the arid environment and the enormous investment involved in constructing a *maʿyān*. Proper maintenance of the system was of utmost importance to guarantee long-term use, and so was fair water allocation, overseen by the *muqaddam* and his advisory team and implemented by the farmers. Successful operation of the canal system required not only social consensus but also consideration for the natural resources, and even though problems with a declining groundwater table did occur and posed a great challenge for the entire community, they could be managed.

12 Postscript

Drastic changes of recent origin imposed from outside the community have left the population of Ghayl Bā Wazīr in a state of helplessness. During the first field season in January 2001, we felt the farmers' increasing sense of despair, and this impression was reinforced during the second field season (December 2003–January 2004). The town is in the province of Ḥaḍramawt, which was part of the People's Democratic Republic of (South) Yemen (PDRY) until the unification of Yemen in 1990. Prior to 1967, Ḥaḍramawt was part of the Protectorate, officially controlled by the British but hardly developed. Independence in 1967 went hand in hand with major political and economic changes. These included nationalization of land as a prerequisite for the reform and modernization

63 Boucharlat 2003, 170.

64 Bagg 2000, 127–146. Muscarella (1986, 468–469) had already pointed out that Laessøe's modification of the text was faulty.

of the agricultural sector.[65] In the post-1967 era, the traditional practices in irrigation agriculture were steadily replaced by so-called modern ones. Yet the nationalization and modernization process ignored that the traditional practices were an integral element of a complex system of regulatory means based on community collaboration and social consensus. While abandonment of that system led to animosities within the community that are felt to this day, the much more severe problem was caused by blatantly neglecting the basic principles of sustainability.

In order to supply the provincial capital, al-Mukallā, with drinking water, in the 1980s government authorities decided to tap the water resources of Ghayl Bā Wazīr, which is located some 40 km to the northeast (see fig. 67). In 1986 the Mukallā Water Project started to pump from five wells. Following Yemen's unification in 1990, al-Mukallā became one of the fastest-growing cities in the country.[66] Therefore, from 1998 onwards, deep-well pumping in the area of Ghayl Bā Wazīr was expanded dramatically to meet the increasing needs of the provincial capital.[67] In addition, in the late 1990s, the Province of Ḥaḍramawt reassigned land that had been nationalized post-1967 to its original owners, and this led to a considerable intensification of farming, including irrigation. As a consequence, between 1998 and 2000 the water table in Ghayl Bā Wazīr dropped 6 m.

At no point were the people of Ghayl Bā Wazīr involved in the decision-making process at the official government level, and they were not in any way consulted to contribute to the policy making with regard to the pumping practices. Current overexploitation of the aquifer has caused the water table to drop rapidly, which is noticeable in the falling water level in the collapse sinks. Between our first field season in January 2001 and the second visit three years later (December 2003–January 2004), we observed a change of several metres. The drinking and domestic water of Ghayl Bā Wazīr is now provided by the Mukallā Water Project, but for the local farmers, the dramatically decreasing groundwater means that the long-established *maʿyān* tradition is coming to an end. They have no choice but to resort to diesel pumps. In January 2001 there was still water in Ḥawmat al-Sirkāl, albeit at a level much lower (some 17 m below ground surface) than prior to the Mukallā Water Project. The farmers were pumping this water back into the existing *maʿyān* (fig. 91). For the same purpose, deep wells were drilled close to those collapse sinks that had dried up. While the method of water extraction differs from the past, the distribution

65 See Pritzkat 1999, 25–31, for details.
66 Ibid.
67 Estimated population in 2003: ca. 300,000.

FIGURE 91 Water pumped from Ḥawmat al-Sirkāl is fed into the existing *maʿyān*

through the *maʿyān* system has remained the same, at least for the time being.

Unfortunately, by January 2004 the water level in Ḥawmat al-Sirkāl had dropped to more than 22 m below ground surface. The farmers realized that given the alarming rate at which the groundwater was decreasing, the use of diesel pumps would offer only a short-term solution. They started a local initiative to improve the situation by diverting the runoff from a nearby wadi and feeding it into one of the collapse sinks as well as into some of the natural joints and fissures (*khuwaʿ*). Based on their understanding of the hydrogeological characteristics of the karst and the underground connectedness of the water, they were hopeful that they would be successful in raising the water table in the area.

There was an expectation on the part of the farmers that we, as educated Westerners with access to funding, would be able to come up with a solution for the problem. When conducting field research in a country that is faced with both extreme poverty and extreme water scarcity, retreating into the glorious past and closing one's eyes to the severe problems of the present can pose an ethical dilemma.

PART IV

Water-Storage Systems in the Western Highlands of Yemen: The Cisterns of al-Jabīn

Introduction to Part IV

The area around al-Jabīn in the western highlands of Yemen offers a unique opportunity to study a different kind of traditional water engineering defined by local characteristics. The modest town is the market and service centre for the district as well as the seat of administration and a military garrison. In 2004 it became the capital of Yemen's newly established Governorate of Rayma, named after the mountain range Jibāl Rayma that dominates it.[1] Al-Jabīn is perched on the rugged western escarpment at an altitude of approximately 2400 m. Due to the steepness of the terrain, arable farming is limited to man-made terraces.[2] The western slopes facing the Red Sea receive significant precipitation during the two rainy seasons in spring and late summer that were described in the Introduction to this book. Additional moisture stems from the diurnal pattern of air movement between the Red Sea and the Yemeni highlands.[3] The flow of warm, humid air from the sea towards the mountains reaches its maximum in the early afternoon. This leads to clouds forming and building up against the escarpment (fig. 92). Al-Jabīn is shrouded in thick fog in the afternoon during much of the year. Condensation of the fog results in dewfall, which provides a daily source of water for farming. Indeed, the conditions for plant cultivation on the higher western escarpment have been called the best in all of Yemen.[4] Today, perennial coffee bushes produce a valuable cash crop. As early as the eighteenth century, the German traveller, cartographer, and surveyor Carsten Niebuhr described the area around al-Jabīn as particularly rich in coffee cultivation.[5]

There is a noticeable difference between the crops grown on the western mountainsides and those on the drier east-facing terraces lying in the rain

1 Prior to 2004, Rayma was part of the Governorate of Ṣanʿāʾ.

2 The origins of terraced agriculture in the highlands extend back to at least the third millennium BCE; see Wilkinson 1999, 185 and 188–190.

3 The phenomenon is described by Munro and Wilkinson (2007, 19).

4 Al-Hubaishi and Müller-Hohenstein 1984, 77.

5 Niebuhr (1772) 1969, 248. See Case Study 1, n. 26, for more information on Niebuhr. He gives the name as "al-Jabī," which is the name that is also listed by Rathjens and von Wissmann (1934, 41). The name change from al-Jabī to al-Jabīn happened quite recently; see al-ʿArūsī 2003, 864. So far, *qāt* (khat [*Catha edulis*, Celastraceae], a shrub whose leaves are chewed by a large part of the Yemeni population every afternoon for their stimulant effects) is grown to a lesser extent on the western terraces. No doubt this will change as the engineering of a new paved road provides the opportunity to deliver *qāt* easily to an external market. See Varisco 2012, 69–70, for more details on *qāt*. Trees that provide hardwood timber are cultivated as well. The timber is often used for roofing beams because of its resistance to termites.

 | DOI:10.1163/9789004387713_012

FIGURE 92 Cluster of houses along the road to al-Jabīn, with terraces and rising clouds in the late morning

shadow (fig. 93). Here, the subsistence crop sorghum provides both cereal grain and animal fodder. Where direct precipitation is insufficient for terraced agriculture, the farmers harvest rainwater—and with it plant nutrients—from the adjacent slopes, which serve as minicatchments. Diversions are created either by digging ditches (where possible) or, more commonly, by heaping up gravel and stones on the open surfaces. As soon as it rains, the surface runoff from the hillsides is funnelled through these small unlined canal-like structures (*sawāqī*, sg. *sāqiya*) onto the lower-lying terraced fields (fig. 94). The individual plots are surrounded by low retaining walls with small openings at their bases to control the runoff and direct it from one terraced field to the next.[6] The runoff collection area has been aptly compared to a roof, and the canal-like structure running along its edge to the gutter collecting the water.[7] The literature refers to rainwater harvesting as "*sawāqī* supplementary irrigation."[8]

6 Rainwater harvesting has been described by Kopp (1981, 110–111) and Eger (1984 and, in more detail, 1987).

7 Hovden 2014, 57.

8 E.g., Kopp 1981, 110–111; Eger 1984, 150; and Eger 1987, 47. Please note that I do not consider *sayl* irrigation to be a method of rainwater harvesting, as suggested by Eger (1984, 150, and 1987, 47), who divides rainwater harvesting into two methods, *sawāqī* supplementary irrigation and *sayl* irrigation. "Harvesting" implies purposeful collection, which is always practised by the farmers in *sawāqī* supplementary irrigation but is not a necessity in *sayl* irrigation,

FIGURE 93 East-facing terraces below al-Jabīn

Because of al-Jabīn's geographical location, digging wells is not an option. Since the town is set at the edge of the western mountain ridge, with the land falling steeply down to the Red Sea, there are no perennial springs located close by. Year-round spring water for drinking is only found at levels much lower down the mountains; springs at higher altitudes and closer to the town are typically seasonal and tend to dry up shortly after the rains. A water-supply

which depends to a large extent on natural flow patterns resulting in the collection of runoff water in a wadi with little or no human intervention. In the same context, Eger's distinction between "overland runoff" and "surface runoff" seems too artificial and arbitrary to be used to define the characteristics of either *sawāqī* supplementary irrigation or *sayl* irrigation.

FIGURE 94 *Sāqiya* for rainwater harvesting

system for al-Jabīn was designed (and started) in the 1980s, and work resumed at the beginning of the 2000s. The system was to be fed by a spring, from which the water was to be pumped into the town. Metal pipes were even installed in the streets, together with a water meter for each house. Unfortunately, due to disputes over the water rights, the system was not completed.[9] Therefore, even today, consumption of spring water involves hauling it uphill in containers carried on the backs of people or draught animals. Nonpotable water for domestic consumption (including washing oneself, doing the dishes and the laundry, and watering animals) and for ritual ablutions comes from open cisterns (*birak*, sg. *birka*) that collect surface runoff following a rain, based on the principles of rainwater harvesting. Al-Jabīn has two large public cisterns, one in the southern and one in the northern part of the town. The latter is referred to locally as Birkat al-Ḍiyāʾ and is fully functional (fig. 95).

The southern cistern, situated below the military fort, is named Birkat ʿĀṭif (fig. 96).[10] It developed cracks several decades ago. Construction of the first road connecting the Tihāma coastal plain with al-Jabīn in the late 1970s and early 1980s was enormously difficult in the rugged mountainous terrain and involved extensive dynamiting. According to the locals, the resulting vibrations

9 See also Swagman 1988, 77, for the first stage.

10 Swagman (ibid., 8 and 100–101) lists al-ʿĀṭif as the name of a tribal descent group settling east of the Jibāl Rayma area, the rugged mountain range where al-Jabīn is located.

FIGURE 95 Children fetching water at Birkat al-Ḍiyā', the public cistern in the northern quarters of al-Jabīn

caused the cistern to crack.[11] Proper repair would have been costly. On the advice of European development aid workers (admittedly from an anthropological project, not a technical one) who visited al-Jabīn in the late 1980s, an attempt was made to mend its lower part with Portland cement—at the time a cheap and readily available building material. However, this turned out to be ineffective, and the cistern retained water only up to a height of 2.5 m, with the rest of the water draining away slowly through the cracks. As Birkat ʿĀṭif remained empty for most of the year, it was possible for us to conduct a technical study of it.

Cisterns are a largely neglected topic in the Western academic literature on Yemen. Spectacular views of the storage reservoirs clinging to the edges of the high mountains are much admired and regularly shown in coffee-table books.[12] However, detailed technical investigations that address such aspects as the construction materials and methods, the catchment area, and the volume of the stored water are mostly lacking. Only very limited information is available on the historical and social context of the cisterns or on issues such as

11 Past and present road construction in the area—the course of the original road was modified in the 2000s—has also led to considerable damage to the coffee terraces on the western mountainsides. This has caused enormous resentment among the farmers.

12 See, for instance, Costa and Vicario 1977, fig. 96; and Maréchaux 1980, figs. 16–19.

FIGURE 96 Birkat ʿĀṭif, with military fort on its north side

community needs and sustainable use. Among the available reports is a typology of Yemeni cisterns that is based on survey work carried out in the late 1920s.[13] Eirik Hovden's 2006 MS thesis provides an anthropological case study of rainwater-harvesting cisterns in the Governorate of Ḥajja (in northwestern Yemen) within the framework of local water-management practices; an informative follow-up article by the same author addresses both the functional

13 Rathjens and von Wissmann 1932, 144–158.

side and the legal framework regarding the ownership and management of cisterns.[14]

At the time of our first field season (December 2007–January 2008), the local population expressed hopes that it might be possible with external help to restore Birkat ʿĀṭif to full use in order to ease the water problems in the town. With one of the large public cisterns being for the most part derelict, women and young girls in particular had to allocate a significant amount of time each day to fetching water. Buying water from water trucks is too expensive for all but a few in al-Jabīn. For us, responding to the request to restore Birkat ʿĀṭif offered an opportunity to give something back to the people who hosted us and generously shared their knowledge not only during the first field season but also during two consecutive field seasons from December 2009 to January 2010 (when the restoration project was in its final planning stage) and from December 2010 to January 2011 (when work was in full swing).

It was the explicit aim of the restoration project to employ the original construction techniques and materials. Restoration work has to start with the removal of layers of previous repair work, and it thus exposes the original structure (fig. 97). Therefore, the project provided us with a unique opportunity to gain insights into technical details that would otherwise have been difficult (or impossible) to assess. Moreover, restoring a cistern to full use entails clearing the runoff collection area, which gives a better understanding of how the cistern was set into the natural landscape in order to make best use of the available water resources. In local parlance, the term *birka* has two meanings. Narrowly applied, it refers to the water-storage reservoir itself, but in a broader sense, *birka* signifies the water-storage reservoir and the entire catchment area with its rainwater-harvesting structures on which the collection of runoff depends; the reservoir itself is no more than part of the system.[15] Clearly, for the locals, the reservoir and its catchment area form an integrated whole.

14 For the MS thesis, see Hovden 2006. The article, Hovden 2014, is based in part on the MS thesis but also incorporates additional material from later field research, including work in al-Jabīn. Hovden was a team member during the second field season in al-Jabīn (December 2009–January 2010).

15 See also Hovden 2006, 53; and Hovden 2014, 57 and 62.

FIGURE 97 Demolition of the Portland-cement layers on Birkat ʿĀṭif

CASE STUDY 7

The Study and Restoration of Birkat ʿĀṭif, a Public Cistern in al-Jabīn

1 Technical Study

Birkat ʿĀṭif collects surface runoff from the surrounding terrain (see fig. 98). The catchment area on the south side is relatively short, with water draining through natural fissures and artificially created diversions from the grounds of a cemetery. The main source of water is rain falling on the roofs and paved courtyard of the military fort on the north side of Birkat ʿĀṭif, from where the water runs approximately 105 m towards the cistern by way of an open canal. Rather than flowing straight into the main reservoir, the surface runoff from both the north and the south sides is fed first into a settling basin to reduce the quantity of solid particles entering the cistern. An overflow chute allows surplus water to be directed away from the reservoir once full storage capacity is reached. The water is channelled onto the terraced fields below the cistern.

Birkat ʿĀṭif is roughly oval in shape, with walls consisting of nine rows of high ledges and a flight of twenty broad steps leading to the bottom (see fig. 99). Many of the steps have one to three small raised "pods" at roughly half a step in height (see fig. 116). These provide easy access to the water at all times, as the water level changes throughout the course of a year. The dimensions of the cistern at its top are approximately 27 × 20 m, and its depth (below the overflow chute) measures 7 m. This gives the cistern a storage capacity of some 1375 m^3. According to local memory, before the cistern cracked and became partly dysfunctional, the water was used by about three hundred families. At the top, three small basins were designed to be filled by bucket as animal-watering troughs or for doing laundry (fig. 100).

The cistern was constructed to fit the chosen site and was built mostly against and into the natural bedrock, with additions of stonemasonry where necessary. As a result of the removal of the Portland cement at the start of the restoration work, it was possible to inspect the composition of the masonry of the lower ledges (fig. 101): it consists of raw fieldstone laid in lime mortar.

A small exposure dug into the floor of the cistern at the base of the bottom step during restoration revealed that the masonry was set on top of a floor that consisted of material similar to concrete. The floor itself was poured 15 cm thick in a single swath, in part over exposed bedrock, with a compacted mud

 | DOI:10.1163/9789004387713_013

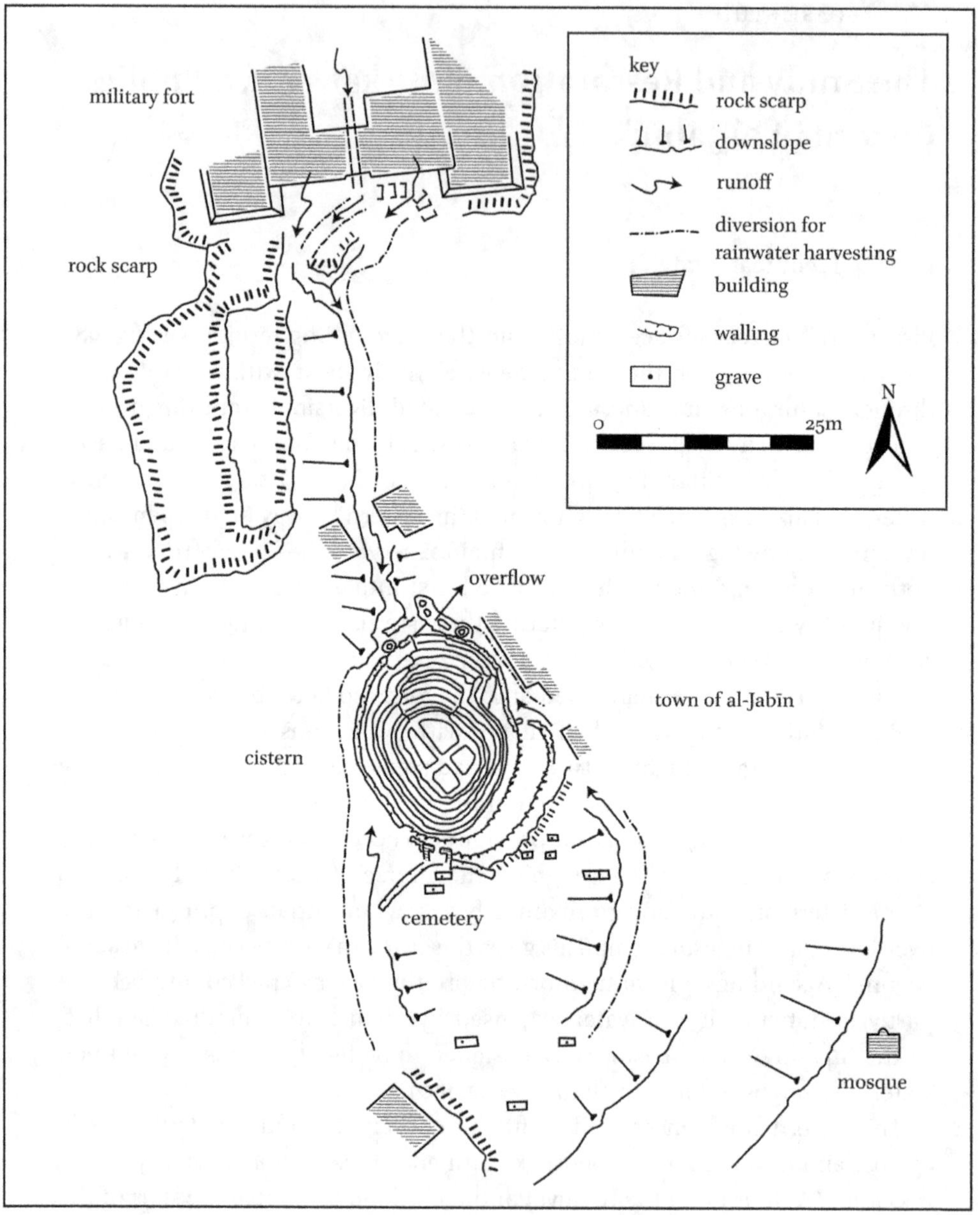

FIGURE 98 Site plan of Birkat ʿĀṭif showing runoff-collection areas

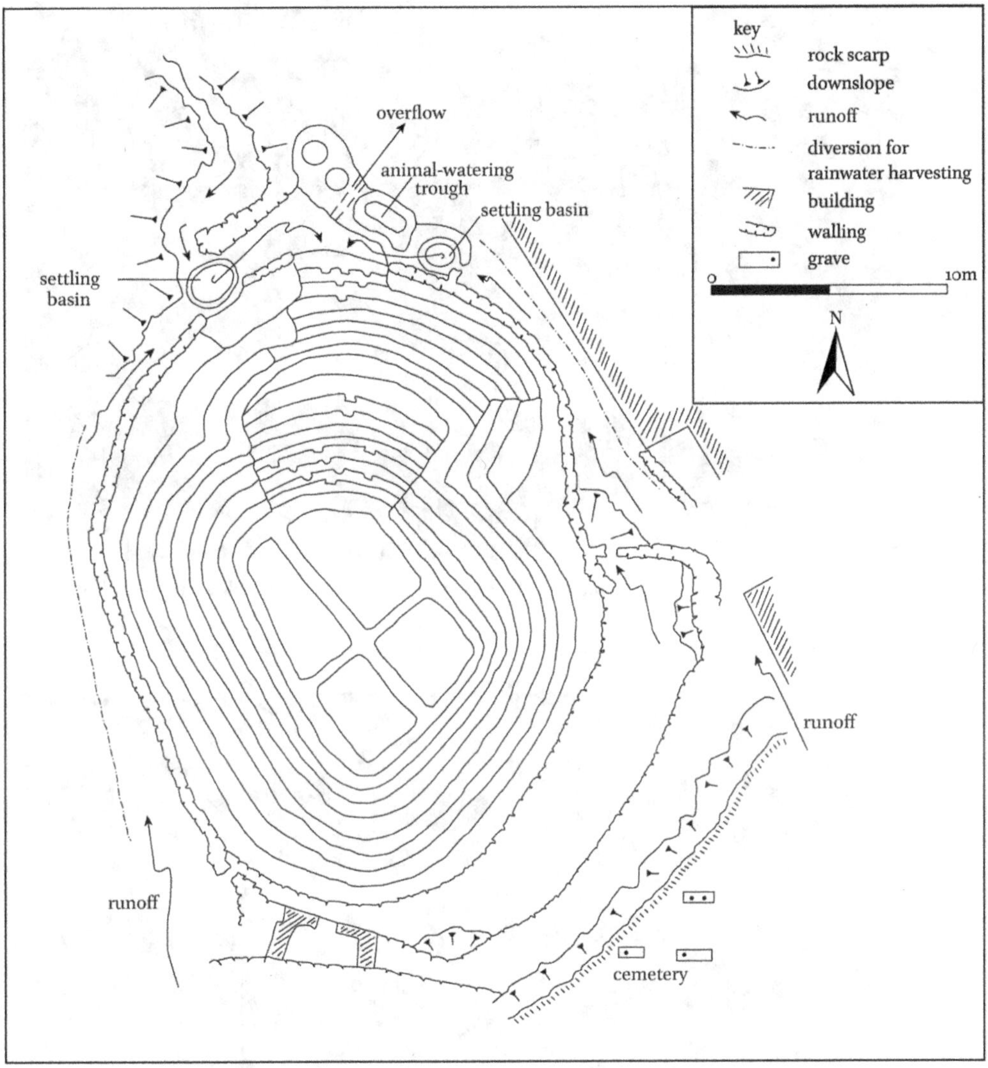

FIGURE 99 Ground plan of Birkat ʿĀṭif

and stone fill (where necessary) to create a level working surface (see fig. 102). There are no signs of floor re-laying or resurfacing, other than the aforementioned recent Portland-cement layer, which also included two crossbars (fieldstone laid in Portland cement) constructed across the floor.

Removal of the Portland-cement layer exposed the original waterproof lining of the cistern (fig. 103). It is composed of three layers. The base layer consists of a moderately fine-grained material, with its surface grooved by pounding marks. The main layer, some 3 cm thick, includes an aggregate of pebbles

FIGURE 100 Animal-watering trough at the top of Birkat ʿĀṭif, with overflow chute (*below*) and settling basin (*right*)

FIGURE 101 The masonry of the two lower ledges exposed: fieldstone laid in lime mortar

FIGURE 102 The floor of the cistern, consisting of a material similar to concrete, laid in part over bedrock

FIGURE 103 The three layers of the original waterproof lining, with remains of the swirling design (*top left*)

(a natural assortment, partly rounded, coming from a stream bed), with a general size of up to 3 × 1.5 × 1 cm. The top surface layer is a thin plaster (no aggregate) with a distinctive finishing touch consisting of a combed swirling design.[1] As conversations with the master builders in December 2010 made clear, waterproofing always involves a succession of layers of different compositions.

The steps of the cistern had been re-laid twice. As a "high-traffic area," they are exposed to considerable stress. This explains why the waterproof lining had required repairs. They were carried out in more or less the same way as described for the main and the top surface layers (no base layer), including aggregate size and assortment.

2 Restoration Work

The following description of the restoration work is grounded in conversations with senior master builder Yaḥyā Muḥsin Ṣāliḥ and junior master builder Muṣṭafā Muḥammad Muḥsin, his nephew, in addition to personal observations made between December 2010 and January 2011. Craftsmen often hand down their know-how within their families, and the senior master builder of the Birkat ʿĀṭif restoration project is a typical example. The term for master builder, not only in al-Jabīn but also in all of Yemen, is *usṭā*, pl. *asāṭiya*, a colloquialism that is derived from *ustādh*, pl. *asātidha*, meaning "teacher" or "professor" and used as a title "when addressing normally older men of higher rank who are considered to possess an expert knowledge or skill."[2] As a rule, a young apprentice works for several years under the guidance of an *usṭā*. After he has finished his training, it takes approximately ten years of working in the profession before he may claim the title of *usṭā* for himself.[3]

Usṭā Yaḥyā and his nephew Muṣṭafā are originally from the Taʿizz area but moved their home base to Ṣanʿāʾ some years ago.[4] Therefore, the terminology that is documented in this chapter may at least in part reflect their provenance. Both were hired because, unfortunately, the local experts from the al-Jabīn area admitted that they had not used traditional construction techniques and materials in decades. They also insisted that the pay for the restoration work should be calculated in silver riyals. The price would be determined by the

1 Hovden (2006, 81 and fig. 28) observed similar decorations on the cisterns in Ḥajja.

2 Marchand 2012, 202.

3 See also Lewcock and Serjeant 1983, 479.

4 In 1986 UNESCO declared the Old City of Ṣanʿāʾ a World Heritage site. Since then, a number of major restoration projects have been carried out there, creating a considerable amount of work for master builders who specialize in traditional building materials and techniques.

number of one-riyal silver coins that could be placed side by side on the surface of the waterproof lining.[5] The silver currency was discontinued in North Yemen after the revolution of 1962.

Our technical study of Birkat ʿĀṭif provided the basis for planning the restoration project. The practical work began in February 2010. In order not to be disrupted by water collecting in the bottom of the cistern following a rain, work started at the top. The senior master builder hired the construction team from the local community. It included two workers who were responsible both for removing those layers of waterproof lining that were judged to be weak and for removing the former repairs. This involved, in particular, demolishing the Portland-cement layers (fig. 97). Each of the master builders also worked with an apprentice to apply the waterproof lining of the cistern, by far the most time-consuming (and costly) step of the restoration work. Figure 101 shows the senior master builder on the left and his junior colleague on the right, with their apprentices working between them. Another man prepared the different waterproof-lining mixtures, and another delivered buckets containing the mixed batch and water to the two teams. The on-site manager's responsibility was ordering the raw materials and paying the bills. Depending on the stage of the work, additional labourers were hired on a short-term basis.[6]

The key component of the waterproof lining is quicklime (*nūra*, chemically speaking, CaO, or calcium oxide). It was bought from lime kilns in Ḥays, a town on the Red Sea coastal plain some 35 km south of Zabīd. In the kilns, burning the raw material limestone (calcium carbonate) results in the following reaction: $CaCO_3 \rightarrow CaO + CO_2$.[7] The gaseous carbon dioxide escapes. To process the quicklime further and prepare the waterproof-lining mixture, three bins were constructed at the top of the cistern. In the first one, lumps of quicklime were mixed with water and left to sit for approximately one week: $CaO + H_2O \rightarrow Ca(OH)_2$. Due to the strong heat development and increase in volume during the forceful chemical reaction between quicklime and water (referred to as "hydration" or "lime slaking"), the lumps of quicklime break up into calcium hydroxide, or slaked lime. In the middle bin, a man in rubber boots stomped on the slaked lime and more water to form a homogenous viscous paste (fig. 104). Any remaining lumps and impurities were eliminated in the

5 To give a specific example, a one-riyal silver coin minted in Ṣanʿāʾ in the year 1344 of the Islamic calendar (1925/6 CE) has a diameter of 3.9 cm and weighs 29.9 g.

6 For instance, when the crossbar at the bottom of the cistern was replaced (see below), a stonemason and his assistant were commissioned to complete the work.

7 See Döring 2012, 293, for a succinct description of the individual steps required to prepare lime as a binder for building materials.

FIGURE 104 Mixing the *nūra* paste through stomping

process. This viscous paste of slaked lime and water, also called *nūra,* is the binder in the traditional waterproof lining.

In the third bin, the *nūra* was mixed with the aggregate, of which three different kinds were used:

- grey grit (*hilsin*)
- smaller-sized gravel (*karrī rubʿ*), measuring some 2 × 1 × 1 cm
- coarse gravel (*karrī niṣf*), measuring up to approximately 3.5 × 2 × 1 cm

The aggregate is required to add strength to what can be technically designated "concrete." Unlike the aggregate in the original waterproof lining of Birkat ʿĀṭif, the gravel used in the restoration project was angular because it was not a natural assortment from a stream bed but was rather made by a crushing machine.

For the base layer (first layer) of the waterproof lining, the aggregate comprises three buckets of grey grit and seven buckets of smaller-sized gravel that were mixed with five buckets of *nūra* (see table 11). The aggregate of the second layer contains two buckets of grey grit and eight buckets of smaller-sized gravel, and was mixed with the same amount of *nūra* as the material for the first layer. The third layer's aggregate comprises one bucket of grey grit, seven buckets of smaller-sized gravel, and two buckets of coarse gravel. It is noticeable that the aggregate is increasingly coarse-grained from the first to the third layer. For the floor, no grit was used, only gravel (five buckets each of the smaller-sized and the coarse gravel), in order to provide the required strength of the concrete. As with the *nūra* paste, the mixing of the *nūra* and the aggregate was accomplished through stomping. The men wore rubber boots as well as rubber gloves when working to avoid the caustic effects of the quicklime. The ratios given are the result of experimentation by the construction team over the course of several months. The original ratios were different but were considered unsatisfactory, in particular because the amount of *nūra* used was judged to be too high. Since *nūra* is expensive, the aggregate ratios were modified in order to stay within the given budget.

TABLE 11 Composition (measured in buckets) of the individual layers of the waterproof lining

	Aggregate			**Binder**
Layer	**Grey grit**	**Smaller-sized gravel**	**Coarse gravel**	***Nūra***
Base layer	3	7	0	5
Second layer	2	8	0	5
Third layer	1	7	2	5
Floor	0	5	5	5

The application of each layer was immediately followed by pounding with the edge of a flat, hand-sized stone to compact the mixture and eliminate voids (fig. 105). The stone used for layer three had a more rounded edge than the one used for the first and second layers (fig. 106). The Arabic term for the pounding stone is the same for both shapes: *mawkhasha*. Due to the small size of the aggregate of the first layer, the mixture could be pounded into the spaces between the (irregular) fieldstone masonry that makes up the ledges of the cistern. The surfaces of layers one and two are grooved by the pounding marks, which allows for a strong bond with the following layer (fig. 105).

FIGURE 105 Pounding of the individual layers of the waterproof lining results in a grooved surface

Layer three was built up in several stages until the desired thickness, flatness of the faces, and horizontality of the step surfaces were achieved. At each stage, the mixture was pounded, and its rough surface was subsequently flattened with a modern metal float (fig. 107). Additional mixture could be added at this point to ensure that the surfaces were flat and level.

The final stages of layer three were smoothing with an oblong riverine (and therefore rounded) pebble (*nāʿim*) (fig. 108) and creating a chamfered edge to the step. This was done while the mixture was still wet (laid the previous day) but usually required sprinkling some water on the surfaces to dampen them to make the matrix more workable. Once even surfaces were achieved and the waterproof-lining mixture had dried, the finished surface of layer three was treated by sanding down any remaining rough protrusions using the flat side of a conical pumice stone (*nashaf*) (fig. 109).

During drying, the slaked lime in the waterproof-lining mixture slowly reacts with carbon dioxide in the air and reverts to calcium carbonate. That is, the binder hardens, a process that is also referred to as "setting": $Ca(OH)_2 + CO_2 \rightarrow CaCO_3 + H_2O$. The aggregate is thus fixed in its position. The water that forms in the reaction evaporates.[8]

8 Döring 2012, 293.

FIGURE 106 The pounding stones (sg. *mawkhasha*) used for layers one and two (*right*) and layer three (*left*)

FIGURE 107 Building up layer three: application of waterproof-lining mixture and flattening with a metal float

FIGURE 108 Smoothing layer three with a rounded riverine pebble (*nāʿim*)

The topcoat, a thin layer of watery lime (no aggregate), was applied with a floor broom made of palm fronds, resulting in a broad but shallow swirling design (fig. 110). Inevitably, hairline cracks developed during the drying process; they were subsequently eliminated through light polishing while sprinkling some water on the surfaces with a broom. The aforementioned rounded riverine pebble (*nāʿim*) was used as a tool. Polishing resulted in the partial disappearance of the swirl design on those surfaces that received the treatment. On exposed edges where considerable work was required, the design disappeared completely. The polished surfaces have a marble-like appearance.[9]

9 See also Al-Radi 1997, 194.

FIGURE 109 Sanding down layer three with a pumice stone (*nashaf*)

About two weeks after completing a ledge, the join between the vertical and horizontal surfaces was filled with a coping (fig. 111). This was necessary as the join is particularly susceptible to developing cracks. The same materials and procedure were used as described above.

At the bottom of the cistern, the existing crossbar was replaced by a more substantial one built with fieldstone set in lime mortar, for which the mixture for the base layer of the waterproof lining was used (fig. 112). Reportedly, after all the surfaces had dried, a coating of rendered animal fat was applied to them in order to penetrate the plaster, fill up its pores, and, as the senior master builder pointed out, "enhance its waterproof qualities." In fact, the chemical reaction between fatty acids and calcium (saponification) results in the

FIGURE 110 Topcoat of watery lime with swirling design

FIGURE 111 Coping between the vertical and horizontal surfaces of two ledges

FIGURE 112 Crossbar at the bottom of the cistern being rebuilt

formation of lime soap, which is water insoluble and seals the surface. As on other construction sites in Yemen, the source of the fat was bone marrow from cattle (from long bones such as femurs). The bone marrow was heated to boiling point prior to being rubbed into the surfaces with a piece of sheepskin, a process that has been described as "extremely smelly and unpleasant."[10]

Each individual step, from applying the waterproof-lining mixture to pounding, flattening, smoothing, polishing, and, finally, rubbing the animal fat into the finished surfaces, needs to be repeated again and again. For instance, eliminating hairline cracks that develop during drying requires painstaking polishing of the surface with a rounded riverine pebble beginning at one end of a ledge and moving slowly forward to the other end, only to start all over as soon as it is reached. Depending on the size of the surface, polishing takes days or even weeks to complete. Application of the waterproof lining "is an expensive, labour-intensive, and time-consuming process that simply cannot be hurried."[11] However, both master builders emphasized that the result is durability.

Instead of harvesting the rainwater through the unlined canal-like structures (*sawāqī*, sg. *sāqiya*) created by digging ditches or by heaping up gravel and stones, the master builders decided to maximize runoff collection by building proper canals, contained within solid revetments (fieldstone set in lime

10 Ibid., 195.
11 Ibid., 191.

mortar) and paved (fig. 113). This measure eliminates loss of water through the natural fissures in the bedrock; it also helps reduce the quantity of solid particles entering the cistern with the runoff and ensures that the inflowing water is as clean as possible. Restoration work was completed in the early summer of 2011, in time for collection of the surface runoff during the rainy season in late summer.

3 *Qaḍāḍ*

The traditional Yemeni plastering material that is applied in multiple layers to a building's external and internal facades as waterproof lining is called *qaḍāḍ*. Lavish decorations are also executed in *qaḍāḍ* by carving.[12] *Qaḍāḍ* has been widely used in Yemen since pre-Islamic times: for instance, on the hydraulic structures in ancient Mārib where it has survived the millennia due to its durability.[13] Several authors have described its composition, preparation, and application, which vary widely. The majority of the publications focus on the Yemeni capital, Ṣanʿāʾ;[14] one focuses on Zabīd;[15] and one in-depth investigation was carried out in the city of Radāʿ, on the highland plateau southeast of Ṣanʿāʾ.[16] Of course, regional differences are not surprising given the variations between locally available raw materials. Yet even within the context of Ṣanʿāʾ, divergences are clearly noticeable. One needs to remember that vertical, horizontal, or dome-shaped surfaces, as well as different purposes (waterproofing or decoration), require specific kinds of *qaḍāḍ* that vary in overall composition and particle size. For instance, on a horizontal surface, the *qaḍāḍ* can be slightly more fluid than on a vertical one, while a finer and perfectly homogenous mixture is needed for decorations. The master builders specializing in *qaḍāḍ* are passionate about their recipes and their individual ways of working. The two components that are always found in *qaḍāḍ* are a binder (*nūra*, the viscous paste of slaked lime and water) and aggregate (from sand to grit to gravel). In the volcanic regions of Yemen, ground volcanic cinder (*ḥashāsh*) is used as aggregate, replacing the aforementioned components either in part or completely.[17]

12 For examples, see ibid., figs. 29, 30, and 57; and Porter 1997, figs. 84–87.

13 See Case Study 1, section 7.

14 E.g., Lewcock and Serjeant 1983, 479–481; Bonnenfant and Bonnenfant-Outrebon 1995, 417–445; Klessing 1995, 72–75; and Sutter 1999, 12–15.

15 Keall 1989b, 28–33.

16 Al-Radi 1994, 6–13; and Al-Radi 1997, 191–196.

17 See also Al-Radi 1997, 191–192.

FIGURE 113 Open canal collecting runoff from the western side of the cemetery, with revetment and paving

In order to understand how the admixture of volcanic cinder modifies the properties of *qaḍāḍ*, one needs to remember that the hardening, or setting, of the binder (and the waterproof-lining mixture) is a reversible process. Over time, the calcium carbonate reacts with water (e.g., from rain or as collected in a cistern) and carbon dioxide (from the air) and forms soluble calcium hydrogen carbonate (also called "calcium bicarbonate"): $CaCO_3 + H_2O + CO_2 \rightarrow Ca(HCO_3)_2$. Calcium hydrogen carbonate is water soluble, which means that the *nūra* slowly decomposes, and with it, the waterproof lining. Volcanic cinder contains silicates, and when used as aggregate, these react with the binder (*nūra*; i.e., $Ca(OH)_2$ and water) and form calcium silicate hydrates (a reaction referred to as "hydratation"). Hydrates are insoluble in water, and therefore the setting is irreversible; *qaḍāḍ* containing volcanic cinder is resistant to weathering.[18]

While one would assume that volcanic cinder should be an indispensable component of *qaḍāḍ*, this is not the case. In many parts of Yemen, volcanic cinder is not locally available, and transporting it to al-Jabīn, for instance, is simply cost prohibitive. The senior master builder acknowledged that in Ṣanʿāʾ, Radāʿ, and several other places, volcanic cinder was considered essential; he also explained that master builders in those areas would find the term *qaḍāḍ* inappropriate when referring to waterproof lining not containing volcanic cinder. He insisted, however, that he had chosen the right materials for al-Jabīn, and that he was experienced and meticulous in applying the waterproof lining and working the surfaces to make them durable.[19] After all, when it comes to *qaḍāḍ*, one should not forget that it has been used very successfully in the past solely based on the practical experience of the master builders. Not even modern science can explain every detail that makes it such an excellent building material, praised, *inter alia*, for its high elasticity and durability.[20]

4 Socio-historical Context

A local tradition claims that the cistern was built by the Turks during the first Ottoman occupation of Yemen (1538–1636); a different version traces its origins to pre-Islamic times. We see here further examples of giving credit for a major engineering scheme to strangers, in the first case, to non-Arabs who are not native to the country (the Ottomans), and in the second case, to people

---

18 See Döring 2012, 293–295. Döring describes the process in the context of ancient Roman construction materials. The admixture of volcanic cinder makes the *qaḍāḍ* "hydraulically setting": it sets through binding water molecules (which form an integral part of the calcium silicate hydrate crystals). This means that *qaḍāḍ* containing volcanic cinder can set under water. Admixture of charcoal has a similar effect to volcanic cinder. It seems that *qaḍāḍ* with a deliberate addition of charcoal was used, for instance, on the north and south sluices of the Mārib dam, as indicated by Vogt (2004a, 90), albeit without his properly understanding the reason behind the admixture. The charcoal pieces have been successfully used by archaeologists for radiocarbon dating of the construction, repairs, and use of the sluices; see Vogt 2005, 510–513.

19 Usṭā Yaḥyā did, indeed, have every reason to emphasize his expertise. He was centrally involved with the construction of the Yemeni pavilion (a typical Yemeni house built with traditional materials and techniques) at the 2000 World's Fair held in Hanover, an accomplishment of which he was understandably very proud.

20 E.g., Klessing 1995, 72; Sutter 1999, 12; and Vogt 2004a, 90.

FIGURE 114 Building inscription on Birkat ʿĀṭif (50 cm scale)

who are non-Muslims (the ancient South Arabians). Either way, "the other" is associated with powers that are required to achieve what seems beyond reach.[21] However, the cistern carries an inscription on its top surface (fig. 114); the writing was more legible from a pencil rubbing (fig. 115).[22]

The inscription states that Birkat ʿĀṭif was completed in the year 1085 of the Islamic calendar (1674/5 CE). This takes us to a time when major parts of

21 See also Case Study 6, section 7.

22 As described in the context of the *waqf* inscription in the Iskandariyya mosque-*madrasa* (Case Study 3, section 1), making a pencil rubbing facilitates reading the text.

FIGURE 115 Making a pencil rubbing of the building inscription

Yemen were under Zaydi rule.[23] Zaydism is a branch of Shiite Islam. The Zaydis claim descent from the Prophet Muḥammad and originally established their imamate in the far north of Yemen towards the end of the ninth century. Their claim to power was based on their spiritual superiority, but that claim was supported by the military strength of the northern tribes. From the beginning, this intertwining led to tensions between the religious values of the Zaydi imams and the tribes' secular values and codes of conduct. Tribespeople expect to be dealt with on their own tribal terms and have proven throughout Yemeni history to be largely resistant to government administration; they prefer to maintain their autonomy to the greatest possible extent. It took several centuries for the Zaydi imamate to grow into a political force in northern Yemen, and the Zaydi rulers' authority continued to be frequently challenged by the tribes. Yet, in the seventeenth century, the Zaydi imams were instrumental in freeing Yemen from Ottoman occupation, and they were "carried to power on the flood of tribal assault upon the resented foreign occupier, a flood [they] unleashed but could barely control."[24] Following the expulsion of the Ottoman Turks, the Zaydi rulers struggled to establish supremacy over all of southwest Arabia, not only Upper Yemen, where their core territories were located in the

23 For details on Zaydism in Yemen, see Stookey 1978, 79–95 and 142–155; and Serjeant 1983a, 68–107. A discussion of Zaydi doctrine can be found in Aziz 2011, 19–27.

24 Serjeant 1983a, 78.

north, but also Lower Yemen—that is, farther south and west. It is during this time that Birkat ʿĀṭif was constructed.

As pointed out by Serjeant, the real difference between Upper and Lower Yemen is not a religious one, Shiite versus Sunni Muslims, even though it may be expressed in religious slogans. The dichotomy is rather between belligerent tribes living in fortified settlements in the northern mountains (and steppes) and villagers lacking the warlike spirit in the lowlands.[25] Al-Jabīn lies in between the two areas. Despite its remote location on the rugged western escarpment, which made it a secure retreat in the past, the local population is nontribal, adheres to the *shāfiʿī* school of Sunni Islam, and has strong ties to the Tihāma coastal plain with its important weekly markets. The recognized leader of the community is called a *shaykh*, pl. *shuyūkh*, and like a tribal *shaykh*, he is responsible, today as in the past, for mediating local affairs and settling disputes involving customary law (*ʿurf*), a significant source of power; he is also paid a settlement fee. Yet, unlike a tribal *shaykh*, under Zaydi rule, he was closely linked with the imams' administrative system, most importantly in his role as supervisor of tax collection (for which he received a certain percentage). While this further increased his local position of power, it means that it would be inappropriate in al-Jabīn to understand the term *shaykh* as implying tribal-chief nomenclature.[26] Al-Jabīn and its immediate environs form a minority enclave on the western escarpment, with tribal Shiite territory to its north, east, and south.[27]

The earliest dated structure we were able to document in al-Jabīn that confirms settlement of the site in the second half of the seventeenth century is inside a small mosque that forms part of the military garrison and houses the tomb of a local saint. His tombstone is dated to the year 1072 of the Muslim calendar (1661/2 CE), a little over a decade before Birkat ʿĀṭif was completed.

---

25 Ibid., 77–78.

26 Swagman 1988, 112–114. For another example of use of the term *shaykh* in nontribal-chief nomenclature, see Case Study 6, n. 22.

27 Today this circumstance is frequently cited by locals as the main reason why they have received limited public funding in the past, in particular prior to 2004, when the area was part of the (Zaydi-ruled) Governorate of Ṣanʿāʾ. The lack of funding resulted in considerably slower development than in the neighbouring areas. Strong tribes simply have a stronger voice than nontribal communities, and the strong tribes are Zaydi. See also Swagman 1988, 160–165.

5 Maintenance

Hydraulic structures in general, and cisterns in particular, require proper maintenance as a prerequisite for successful long-term use. This is not limited to the storage reservoir itself. It is important to apply the term *birka* in the broader sense of its meaning because maintenance starts with keeping the water-harvesting areas and canals clean and in good repair to maximize collection of the runoff and ensure that the water is as free of contaminants as possible. In addition, the runoff is fed into a settling basin before entering the water-storage reservoir. Yet sediments inevitably settle at the bottom of the cistern. The annual cleaning should be completed in time for the first rainy season in the spring or, if there has been out-of-season precipitation, as soon as the cistern is empty. A cistern is also carefully inspected once a year. Any hairline cracks that may have developed are treated with watery lime and polished with a pebble. This can be followed by the application of animal fat to the surface. It is of vital importance to carry out such minor repairs as well as any major repairs with the appropriate building materials and techniques. As the example of Birkat ʿĀṭif has shown, using Portland cement on *qaḍāḍ* is not effective because these two materials simply do not bond. One needs to remember that "cement has caused more damage than good when it has been used in old buildings."[28]

Effective maintenance work is not exclusively a technical undertaking, though, and it relies in a major way on social factors. This basic truth can be illustrated through a comparison between Birkat al-Ḍiyāʾ in the northern quarters of al-Jabīn and Birkat ʿĀṭif in the south. Maintenance work on the two cisterns is organized—and, if need be, reinforced—by the two neighbourhood headmen (sg. *ʿāqil*). In the southern district of the town, where Birkat ʿĀṭif is located, this mode of operation functions smoothly, and everybody contributes to the work. To give an example, while we were conducting the technical study in preparation for the restoration project, we needed to sweep Birkat ʿĀṭif clean. No sooner had we started than children from the surrounding houses arrived with brooms in order to help (fig. 116). They pointed out that cleaning was their task, not ours. Clearly, the spirit of shared responsibility is still alive in the southern community. While Birkat al-Ḍiyāʾ in the northern quarters is fully functional from an engineering point of view, it is in an appalling state because it has not been cleaned in quite some time. The people using the water complain about its poor quality but do not appear to take any action towards improving it. As was pointed out by the *shaykh* of al-Jabīn (who lives

28 Al-Radi 1997, 197.

FIGURE 116 Birkat ʿĀṭif before restoration, with crossbar at the bottom and children sweeping

in the southern district), the difference in the northern quarters is the greater number of newcomers from the countryside who recognize neither the *ʿāqil*'s nor the *shaykh*'s authority. The general lack of community cohesion manifests itself in the fact that, instead of participating in the work, the neighbourhood residents normally contribute to the cost of hiring someone to clean the cistern. However, the money seems to have a tendency to leak away.

We can also interpret the situation as an example of how a post-revolution shift in power and wealth has resulted in the loss of the authority of the old political and economic forces. In particular, with the demise of the Zaydi imams, the *shuyūkh* in al-Jabīn and its immediate environs lost their responsibility for tax collection and the profit they made from it; in addition, the new state administrative and judicial systems with their educated professionals offered alternatives for dispute settlement. Since the *shuyūkh* are not tribal leaders, the key tribal value of unity on which they could claim the role of mediator is lacking. As a result, parts of the population of Jibāl Rayma no longer respect a *shaykh* (or an *ʿāqil*).[29]

29 See Swagman 1988, 115–118.

6 Legal Considerations

Communal maintenance work "is also a strong confirmation of the communal status of the cistern."[30] Birkat ʿĀṭif and Birkat al-Ḍiyāʾ are both public cisterns, and their water is shared by their respective neighbourhoods. While irrigation from the cisterns is not permitted, every family has the right to use the water for domestic purposes, as long as the supplies last. Written agreements concerning rights and regulations do not exist in al-Jabīn because they are not needed. The water is ultimately derived from rain and as such is water in its natural state, which in Islamic law is *mubāḥ*: it cannot be owned, and therefore access to it is free to everybody.[31]

A cistern is dependent on its rainwater-harvesting area, without which it cannot function. Therefore, according to customary law, a cistern has the right to the natural runoff feeding it. This also applies where the collection area is not owned by the cistern's owner(s). Birkat ʿĀṭif, for example, has the right to collect the surface runoff from both the grounds of the cemetery on its south side and the roofs and courtyard of the military fort on its north side. The owner(s) of a cistern must give consent for any modifications of the water-harvesting area that result in a decrease of the runoff feeding the reservoir; otherwise, compensation is required.[32] If part of the rainwater-harvesting area is sold, the new owner of the land does not have the right to its surface runoff; that is, he is not entitled to collect even a part of it to feed his own cistern. Needless to say, these issues are frequently contested. The matter of compensation is particularly tricky since money does not solve it. A significant number of cases at the Appeals Court in the Governorate of Rayma (of which al-Jabīn is the capital) are disputes over runoff rights. The water-storage reservoir and the rainwater-harvesting area on which it depends form an integrated whole not only in practical but also in legal terms.

30 Hovden 2014, 61.

31 Maktari 1971, 13–14; see also Case Study 2, section 3.1. For the same reason, those (few) residents of al-Jabīn who have their own private cisterns (which are usually built underground, beneath the courtyard of the house) to collect the runoff from the rooftops sometimes supply this water to their neighbours for only a small fee to cover its delivery by donkey. They point out that the water itself is free because it is *mubāḥ*; one cannot sell something that one does not own.

32 See Hovden 2014, 62–63. For an interesting example of a clash between the right to runoff and the right to land in the Zaydi north of Yemen, see Lichtenthäler 2000, 147–150.

7 Conclusion

Birkat ʿĀṭif is a typical example of the "bold, free-form, rain-fed and rock-cut cisterns …, which are … found on hilly sites [in Yemen]."[33] These have been cited as having possibly exerted an influence on hydraulic structures in other parts of the world, such as the cisterns in Rajasthan. American architectural photographer Morna Livingston gives a detailed description of the cistern of Nahargarh, which was built by the ruler Jai Singh in 1732 and functioned on the principles of rainwater harvesting. Jai Singh was well known for his admiration for geometry. All traditional Indian hydraulic structures follow, in fact, a symmetrical plan with the single exception of the cistern of Nahargarh. Its asymmetrical organic shape is a consequence of its being built to fit the chosen site and to maximize runoff collection.[34] Instead of featuring a symmetrical layout imposed on the landscape, the cistern of Nahargarh reflects a sensitive approach to the natural environment. This is what Livingston interprets as a possible Yemeni influence.

Cisterns collect the surface runoff that flows down mountain slopes following a rain. These systems make use of renewable water resources and are environmentally sustainable. In view of Yemen's current water crisis, which is due mainly to the overexploitation of nonrenewable water resources, preserving the country's numerous cisterns could, cumulatively, have considerable impact on dealing with the problem. The storage reservoirs are an effective means for providing households with supplies of water for domestic consumption. As long as the cisterns are maintained properly, the water is perfectly adequate for this purpose.

The restoration of Birkat ʿĀṭif was generously sponsored by the Social Fund for Development, a Yemen-based funding agency that, *inter alia*, "supports development opportunities through improving access to basic services."[35] "Heritage" and "water" are the top two areas of intervention, with special emphasis on traditional systems that use renewable resources—for example, rainwater harvesting.[36] The restoration project also matched the fourth area of intervention, "training." As mentioned, the master builders from the al-Jabīn area had not worked with *qaḍāḍ* in decades. This is not unusual because Portland cement became readily available after the revolution in North Yemen in 1962 and the opening of the country towards the outside world. Portland

33 Livingston 2002, 130.

34 Ibid., 126–130.

35 Social Fund for Development, n.d., "Background and Objectives."

36 Ibid., "Areas of Interventions [*sic*]."

cement is cheap and allows simple and quick processing, and hence it rapidly replaced *qaḍāḍ* from the late 1960s onwards. In the 1980s, when attempts were made in some restoration projects to go back to using the traditional building materials and techniques on prominent buildings such as mosques or houses in the Old City of Ṣanʿāʾ, master builders who were thoroughly familiar with the details of working with *qaḍāḍ* were initially difficult to find. In most projects, the ratios of the ingredients, the preparation of the mixture, its application, and the individual steps involved in working the surfaces had to be determined through experimentation. Since then, new generations of master builders have been trained, and a building tradition that forms such an important (and aesthetically pleasing) part of Yemen's cultural heritage, one that seemed to be almost lost, has been revived.[37] Moreover, the knowledge gained in various restoration projects is now well documented.

37 Bonnenfant and Bonnenfant-Outrebon 1995, 417; Klessing 1995, 72; Al-Radi 1997, 191 and 197–198; and Sutter 1999, 14–15.

PART V

Water and Religious Magic

Introduction to Part V

Water is a scarce resource in large parts of Yemen. Agriculture requires irrigation to meet the crops' demand for water, and settlements need to be supplied with water year-round. Water engineering—for example, the construction of cisterns for storage during the dry seasons, as documented in al-Jabīn—is an important step in addressing the situation. However, it does not provide the definitive solution as the rains, essential for operation of the engineered systems, are not completely reliable. At irregular intervals, severe droughts jeopardize successful farming and threaten the very survival of communities. Therefore, people did not rely exclusively on their engineering skills but also resorted to practices of a nontechnical nature, two of which will be discussed in this chapter. The first takes us back to premodern al-Jabīn, and the second to ancient Mārib. Their underlying patterns are similar, which emphasizes once more the continuity of practices of water management.

 | DOI:10.1163/9789004387713_014

CASE STUDY 8

Water and Religious Magic

1 The Cistern of Bayt al-Shaykh and the Group of Seven Signs

Bayt al-Shaykh is a cluster of dwellings that today forms the southernmost part of al-Jabīn. It is located south of Birkat ʿĀṭif and next to the Grand Mosque. The houses are centred around a small cistern used exclusively by the settlement (fig. 117). An underground passage connects the buildings of Bayt al-Shaykh to each other and provides access to the cistern. Such tight clusters of dwellings, often built wall to wall to give a fortress-like impression, are typical of the western highlands. The houses usually share a mosque and a cistern, and the inhabitants cooperate in the care and maintenance of both.[1] In Bayt al-Shaykh, the house rooftops and even part of the roofs of the Grand Mosque serve as collection surfaces for rainwater to feed the cistern. In turn, any overflow from it is conveyed into a small pool that is part of the mosque complex and provides water for performing the ritual ablutions before prayer.[2]

The Bayt al-Shaykh cistern's enclosing wall is formed partly by the adjoining houses and partly by the mosque complex. As shown in figures 118 and 119, the plaster surface of the wall bears (from right to left) a six-pointed star (*A*), the *shahāda* (the Muslim creed of the oneness of God and acceptance of Muḥammad as God's prophet; *B*), a group of seven signs (*C*), and a building inscription (*D*).

The group of seven signs is a common motif throughout the Islamic world. A detailed description of it is given by Aḥmad ibn ʿAlī al-Būnī (d. 1225 or 1232/3),[3] a prolific thirteenth-century writer who enjoys great popularity in the Middle East to this day. The main work attributed to him is *Shams al-maʿārif wa-laṭāʾif al-ʿawārif*. However, references to individuals who lived after al-Būnī's death mean that parts of the text cannot have been written by al-Būnī himself, and issues of style suggest the contributions of several generations of authors. Quite understandably, scholars have described al-Būnī's work as difficult, even confusing. Its appearance in different versions adds to the problem. The extant medieval manuscript texts are fairly consistent and much shorter than the printed editions with the title *Shams al-maʿārif al-kubrā wa-laṭāʾif al-ʿawārif*,

1 Swagman 1988, 27–28.

2 For details on the mosque and its ablution pool, see Hehmeyer 2017, 162–163.

3 Gardiner 2012, 89.

FIGURE 117 Cistern of Bayt al-Shaykh, with entrance to underground passage. The inscriptions and the group of seven signs are on the wall to the left of the passage (below the windows)

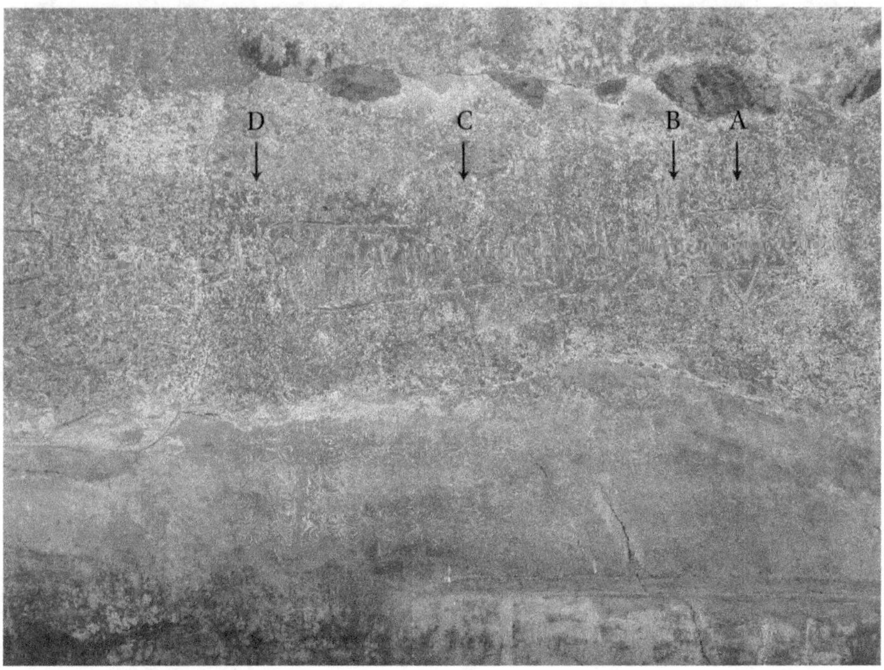

FIGURE 118 Close-up of fig. 117. *Right to left*: six-pointed star (*A*), *shahāda* (*B*), group of seven signs (*C*), and building inscription (*D*)

FIGURE 119 The group of seven signs enhanced via tracing

which have been published since the middle of the nineteenth century.[4] The text that the printed editions reproduce retains certain parts of the medieval version, but it has been argued that it is a compilation that is to a large extent a product of the early seventeenth century. This considerably later date does not affect the importance of *Shams al-maʿārif al-kubrā* as a source for the study of magic. One simply needs to be careful not to attribute the work as a whole to al-Būnī.[5] For the purposes of this chapter, the questions of authorship and the precise date of the different versions of *Shams al-maʿārif* are not significant. For reasons of simplicity, I will also refer in the following to al-Būnī as the author of both the medieval manuscripts and the printed editions.

In a late-thirteenth or fourteenth-century manuscript copy of *Shams al-maʿārif wa-laṭāʾif al-ʿawārif*, al-Būnī gives the following description of the seven signs:[6]

4 El-Gawhary 1968, 14–26; Witkam 2007, 183–187; and Gardiner 2012, 101–105.

5 Gardiner 2012, 123–129.

6 The passage is reproduced in Arabic together with a German translation in Winkler 1930, 68–72, especially 69 and 71. Anawati (1967, 27) provides a French translation. I am following the English translation of the passage by Porter (1998, 145–146), slightly amended. Please note: Porter erroneously cites the theologian Fakhr al-Dīn al-Rāzī (d. 1209) as the author

> Three sticks are lined up after a seal, at their head something that is like the bent head of a lance; a *mīm* squashed and amputated; then a ladder which leads to every hoped for object but which is nonetheless not a ladder; four objects resembling fingers have been lined up, they point towards good things but [they are] without a fist; a *hāʾ* in half; then a *wāw* bent over like a tube of a cupper but which is not a cupping glass.

The signs on the Bayt al-Shaykh cistern closely match al-Būnī's account, except that the three vertical lines lack the horizontal wavy line on top, the letter *wāw* more closely resembles a *mīm*, and the order of the *wāw* and the *hāʾ* is reversed. Of course, Arabic inscriptions are known for the rather liberal use of letter shapes and even the occasional transposition of whole words. As for the seven signs in particular, differences of shape and order were typical until al-Būnī's time, after which only minor variations occur.[7]

1.1 *Origin and Interpretation of the Group of Seven Signs*

It is not a coincidence that the group consists of seven signs. With its origins in the ancient Near East, the symbolism of the number seven has retained great importance in Islamic tradition (as well as in Christianity, Judaism, and other religions). While seven is connected with periodicity, from the seven days of the week to the seven stages of human life, it is also a means of expressing great magnitude. At the same time, seven is considered the number of completion. Moreover, it is a holy number and can therefore be used to avert misfortune.[8]

The seven signs' origin and interpretation have been traced in particular by Hans Winkler.[9] The following remarks draw on aspects of his study that allow us to appreciate why the signs were inscribed on the cistern's perimeter wall. A significant source for Winkler is chapter 12 of *Shams al-maʿārif al-kubrā*. Here, al-Būnī lists the seven signs twice and describes their meaning in considerable detail.[10] Yet the explanations are not always straightforward and are at times even contradictory.

---

instead of al-Būnī. See Gardiner 2012, 103, for the date of the manuscript BnF MS arabe 2647 used for all three translations: "The codex lacks a dated colophon, but the Baron de Slane estimated that it is from the late seventh/thirteenth century, and it certainly is no more recent than the eighth/fourteenth century."

7 Winkler 1930, 67 and 114–119.

8 For these and other aspects of seven symbolism, see Conrad 1988, 43–53; Hartmann-Schmitz (1989, 121–124) provides an in-depth analysis in Sunni Islam.

9 Winkler 1930, 114–149.

10 Al-Būnī, n.d., 1:80–86. See Winkler 1930, 72–85, for a German translation of the critical passage from chapter 12.

Al-Būnī was not the first author writing on the topic. He relied heavily on an earlier work known as the *Dīwān of ʿAlī*, a collection of poetry and sayings attributed to ʿAlī ibn Abī Ṭālib (d. 661), the cousin and son-in-law of the Prophet Muḥammad.[11] It contains a poem about ʿAlī's finding a group of seven signs inscribed on a rock that is identical—word for word—to al-Būnī's account of the first four signs. The remaining signs differ in order and shape.

Heinrich Ewald, who published the *Dīwān of ʿAlī* in 1839,[12] subsequently explained that although the title attributes the work to ʿAlī, its style points to its having been written after his lifetime,[13] a critical observation first made by Joseph Reinaud.[14] Winkler reexamined the poem about the seven signs and noted that the most recent name given in its chain of transmitters, Abū ʿAlī al-Ṭabrisī (d. 1154), suggests the text dates to the first half of the twelfth century.[15] The reference to ʿAlī ibn Abī Ṭālib as the person who discovered the signs should be understood as emphasizing their venerable age and thus their importance. Even though ʿAlī is most probably not the poem's author, the poem's significance through the centuries, especially for Shiites, is underlined by its inclusion in a standard reference work on Shiite Islam.[16]

Ewald published the poem with the title "A Himyaritic Inscription."[17] The Himyarites were the last of the ancient South Arabian dynasties; they conquered the Sabaean and neighbouring realms and unified South Arabia under their hegemony from the end of the third century CE to the Abyssinian conquest in 525. The Himyarites retained Ancient South Arabian for formal writing. The study of those languages had only just started when Ewald was working on his article. This explains why he misinterpreted the seven signs as Ancient South Arabian writing. Another example underlining how little the seven signs were understood at the time is an 1864 account of an engraved stone bearing the names of the twelve imams (of the Twelver branch of Shiite Islam) around the rim and the seven signs at the centre. The author of the article, Andreas Mordtmann, admits that he is not familiar with their meaning, and in the accompanying illustration the seven signs are, in fact, shown upside down.[18]

Unfortunately, it is not possible to date the stone with certainty. However, it shares a peculiarity with some other illustrations of the signs—namely, that the first sign has the shape of a small circle. These examples include the (so

11 Cod. Brit. Mus. 577 [Add. 7534].
12 Ewald 1839a, 107–109.
13 Ewald 1839b, 192–200.
14 Reinaud 1828, 2:245.
15 Winkler 1930, 65–66.
16 See Āghā Buzurg al-Ṭihrānī 1938/9, 3:203–204.
17 Ewald 1839a, 107–109.
18 Mordtmann 1864, 51, no. 13, and plate 6, no. 14.

far) oldest datable piece: a talisman most probably from the eleventh century.[19] The earliest text with an illustration of the group of the seven signs, the poem attributed to ʿAlī ibn Abī Ṭālib, shows the same characteristic. For an explanation of the first sign's significance, we have to go back to al-Būnī's text.[20]

1.1.1 The First Sign: The Seal

As noted above, in the late-thirteenth or fourteenth-century manuscript copy of *Shams al-maʿārif*, al-Būnī calls the first sign simply a "seal." In chapter 12 of the printed *Shams al-maʿārif al-kubrā*, he refers to it as a *hāʾ*.[21] When standing on its own, the letter *hāʾ* of the Arabic alphabet has the shape of a small circle. In fact, this is how the first sign is illustrated in the earliest text examples and on the oldest objects that feature the group of seven signs. In the same text, however, al-Būnī also calls the first sign "a seal with five corners"—that is, a five-pointed star.[22]

This symbol, a five-pointed star drawn in one continuous line, has its own ancient origins, independent of the group of seven signs. The earliest evidence is found in the eastern Mediterranean before 1500 BCE; several examples from the first millennium BCE also exist.[23] Wherever the five-pointed star was used—whether in the eastern Mediterranean or in those territories that adopted it later—the original pentagonal shape was eventually transformed into a six-pointed star composed of two overlapping equilateral triangles. Examples of the latter appear as early as the first millennium BCE and increasingly in the first millennium CE, and both shapes were used in parallel.[24]

As part of the group of seven signs, the six-pointed star is shown in the aforementioned late-thirteenth or fourteenth-century manuscript copy of al-Būnī's work, while in a sixteenth-century manuscript copy and the printed version,

19 See Winkler 1930, 56, no. 1; 114, no. 1; and 116, no. 1. The order of the signs differs slightly from the one described by al-Būnī, with the first sign taking the second position.

20 In a previous publication (Hehmeyer 2008, 87–90), I give a detailed description of each of the seven signs. It is essential to repeat some of those remarks here (with modifications) in order to understand the group's overall significance in the context of this chapter.

21 Al-Būnī, n.d., 1:86.

22 Ibid., 1:84. Please note that in this instance al-Būnī describes the five-pointed star as being repeated at the end of the group. See Winkler 1930, 118, no. 15, for an explanation; an illustration of an example can be found in Doutté 1909, 164.

23 Bliss and Macalister 1902, 83 with plate 29, no. 42; and 122–123 with plate 56, nos. 44 and 53. See also de Genouillac 1926, 115 with plate 20, no. 10043.

24 The five- and, particularly, the six-pointed stars are recognized in popular Islam as the "seal of Solomon," which is associated with supernatural powers. See Milstein, n.d., 186–169 [*sic*]; and Winkler 1930, 57–65, 95, and 127–133. Milstein's catalogue and Winkler (121–126) present examples from ancient and medieval Jewish, Christian, and Islamic contexts.

the five-pointed star is found. Other authors present the seal in its five- and six-pointed shape within the same manuscript. A comparison of the textual descriptions and objects featuring the group of seven signs shows the historical sequence to be as follows: the small circle is the original shape of the first sign (the "seal"); it eventually starts to be replaced with the five-pointed star, and soon after that it becomes interchangeable with the six-pointed star.[25]

The first sign's significance is explained by its original circular shape, together with al-Būnī's statement that it represents a *hāʾ*. This letter is to be understood as an abbreviation for *huwa*, literally meaning "He," the Koranic designation for Allah.[26] As noted earlier, al-Būnī also refers to the first sign as a "seal" (*khātam*). Even today, it is customary in the Middle East to use a seal as a signature, a person's mark that validates such documents as official letters or contracts. Stamped onto an object, a seal establishes authenticity and serves as a sign of ownership.

These functions are confirmed by the material culture remains from Yemen that carry five- and/or six-pointed stars—not as part of the group of seven, but standing on their own. Five- and/or six-pointed stars are found on objects such as coins, where they attest to the item's validity and legitimacy. In its five-pointed shape, the star first appears on coins from ancient South Arabia;[27] in its six-pointed shape, it is found, for instance, on coins from Rasulid Yemen.[28] More importantly for this study, the symbols frequently occur in contexts where they have a protective and talismanic function, with "talisman" being used here in its most general sense as "a thing endowed with potency."[29] In Islamic Yemen, five- and six-pointed stars as geometric shapes are widespread in domestic architecture, on house facades, interior walls, doors, and windows (fig. 120). They also appear in religious contexts—for instance, as part of inscriptions of a religious nature and on religious architecture, such as mosques, minarets, tombs, and cenotaphs. They are used as designs on jewellery and amulets,[30] and as graffiti on walls and rocks.[31] The objects or persons bearing the seal are

25 See Winkler 1930, 114–119.

26 Ibid., 135.

27 Müller 1899, 78, n. 1, no. 15.

28 Nützel 1892, 149–150, no. 85.

29 Savage-Smith 2004, xxiii.

30 Savage-Smith (1997, 133) defines an amulet broadly as "any relatively small object intended to be worn to ensure protection and well-being."

31 For numerous examples from Yemen through the centuries, see the following works: Bonnenfant 1995, especially 537–544, 546–547, and 548–549; Bonnenfant 2004, especially 134–138 and 152–155; Bonnenfant and Bonnenfant 1987, especially 64 (F47), 126 (H1), 169 (139 and 143), and 173 (147); Giunta 2002, especially fig. 4; Golvin and Fromont 1984, 74, fig. 23, and 154, fig. 57, no. 9; Ransom 2014, nos. 130 and 193; Wöhrlin 1999, especially nos. 26

FIGURE 120 Gypsum tracery with five-pointed star on an arch window, al-Jabīn

"stamped" with the abbreviated or symbolized form of God's name and thus "sealed" against, or protected from, harmful influences.[32] Clearly, for Islamic times, the emphasis is on the protective and talismanic nature of the symbols' distinct supernatural qualities, and their appearance in religious contexts is notable.

and 37. The five- and/or six-pointed stars are also part of various practices attempting to influence the course of events by warding off evil. In this function, the stars, either standing on their own or as part of the group of seven signs, are found on what are called "magic-medicinal bowls"; see Canova 1995, especially nos. 2, 3, and 6; and Regourd 2007, 309–345. These objects form a topic that deserves a separate study in its own right.

32 See Winkler 1930, 112.

1.1.2 The Second Sign: Three Vertical Lines with a Wavy Horizontal Line on Top

The second sign, "three sticks lined up …, at their head, something that is like the bent head of a lance," defies explanation.

1.1.3 The Third Sign: The Letter *Mīm*

The significance of the third sign, the letter *mīm*, is derived from the fact that it is the first letter of Muḥammad's name and is repeated in the middle of the word.[33] At the same time, the *mīm*'s numerical value is forty, another important number in Islam (as well as in Christianity and Judaism). "Forty" is used not only in the literal sense of its numerical equivalent but also in a metaphorical way to imply "numerous." This meaning causes it to play a significant role in Muslim folk belief and saint worship.[34] An interesting article by Lawrence Conrad discusses aspects of forty's symbolic value as regards measuring time and describes it as the ideal age for the "optimum balance of physical strength, emotional maturity, and intellectual vigour."[35] Thus, it may also imply spiritual consummation: Muḥammad was "forty" when he received the first revelation.

However, al-Būnī's remark that the *mīm* is "squashed and amputated" remains obscure.

1.1.4 The Fourth Sign: The Ladder

Al-Būnī's description of the fourth sign as "a ladder which leads to every hoped for object" seems plausible, given that its shape does, in fact, resemble a ladder with rungs. The ladder may therefore be understood as a symbol for climbing up to—that is, getting closer to—the good and thus to God. This is why prayer is sometimes called a "ladder." Al-Būnī's remark that the sign "is nonetheless not a ladder" could be viewed as emphasizing its symbolic nature.[36]

33 Ibid., 148.

34 Hasluck 1929, 2:391–402; König 1907, 913–917; and Kriss and Kriss-Heinrich 1960–1962, 1:117, 220–221, and passim.

35 Conrad 1987, 232.

36 Winkler (1930, 148–149) speculates that both the *mīm* and the ladder may have their origin in pre-Islamic signs. He points out that they bear some resemblance to a pair of symbols frequently found on ancient South Arabian monuments. Numerous examples are given in Grohmann 1914, 19–32, where the two signs are referred to as *Blitzbündel* and *Doppelgriffel* and interpreted as symbols of the divine. However, von Wissmann (1982, 76) readdresses the issue and explains the signs instead as a monogram of the ruling dynasty; Müller (1985, 651) also endorses von Wissmann's interpretation.

1.1.5 The Fifth Sign: Four Vertical Lines

Al-Būnī describes the fifth sign as follows: "four objects resembling fingers have been lined up, pointing towards good things but [they are] without a fist." Throughout human history, frightening demons and spirits with only four fingers have populated many cultures.[37] Particularly interesting in the Islamic context is the description of the *jinn* (spirits) of the Punjab, who "are believed to have no bones in their arms; they have only four fingers and no thumbs."[38] The fourth sign would thus be a symbol that provides protection from those fearsome beings with incomplete hands. But in *Shams al-maʿārif al-kubrā*, al-Būnī gives an alternative explanation: he calls the four vertical lines "four *alifs*."[39] The letter *alif* is of special significance because it is the first letter of the word Allah and can thus serve as a symbol for the name of God.[40]

1.1.6 The Sixth and Seventh Signs: The Letters *Hāʾ* and *Wāw*

The letters *hāʾ* and *wāw* combine to form *huwa* (literally, "He," the Koranic designation for Allah). Al-Būnī's description of the "*hāʾ* in half" refers to the letter's shape when connected to the left, in this case to the *wāw*. Occasionally, the *hāʾ* is replaced with a six-pointed star, which seems to underline the aforementioned interchangeability of *hāʾ*, the small circle, and the six-pointed star.[41]

1.2 *The Significance of the Group of Seven Signs*

Subsequent to the description of the seven individual signs, in the late-thirteenth or fourteenth-century manuscript copy of *Shams al-maʿārif*, al-Būnī sums up the meaning of the group as a whole: "This is the name which is supreme in its power, and if you did not know this before, know it [now].... Here is the name of Allāh may his glory be exalted...."[42] As for the practical uses of this group of seven signs in everyday life, he claims that no ship carrying a person bearing them will sink, no house containing them will be destroyed by fire, and no goods marked with them will be stolen. The seven signs promote recovery from sickness and, if written down and deposited with a corpse, will protect the deceased from the torments of the grave. In addition, they will

37 Winkler 1930, 138–140.

38 Hussain 1891, 103, no. 678.

39 Al-Būnī, n.d., 1:85.

40 Winkler 1930, 140. As so often, the interpretation is not necessarily as straightforward as it may seem at first sight. For a less favourable comment on the *alif*, see Goldziher 1872, 782–784.

41 Winkler 1930, 144 and 116, no. 16. For the further development of the six-pointed star into an octagon, see 144 and 116, nos. 17–18.

42 Arabic text in Winkler 1930, 69; English translation in Porter 1998, 146. See also n. 6.

ensure strength and support for those in need of help. These and many other examples listed by al-Būnī highlight their protective and talismanic powers.[43] The underlying notion is to influence the course of events by using supernatural force.

The presupposition of hidden supernatural forces that can be controlled by humans to their advantage through appropriate practices is what defines, in very broad terms, magic.[44] In Islam, magic (*siḥr*) is a well-known and contentious issue and has been addressed by scholars from the seventh century onwards.[45] A clear distinction is made between white magic, which causes no harm to others but rather tries to ward off hardship, and black magic, which seeks to harm others. The latter takes recourse to demonical force and is prohibited, whereas the former is permitted in certain cases. But the division between the two is not as clear as it may seem, for books on white magic normally contain explicit instructions on how to inflict misfortune on others. Al-Būnī's own work can serve as an example of this practice.[46] Black magic will not be further addressed here.

White magic (*sīmiyā'*) is deeply interwoven with religion. At the core of this union lie the *khawāṣṣ*, sg. *khāṣṣa*, the special properties—that is, unaccountable, hidden powers—of such things as Koranic verses and God's names, including their constituents, most importantly the letters of the alphabet.[47] Medieval authors themselves classified use of these secret powers as a practice within white magic.[48] The *khawāṣṣ* take us back to the group of seven signs. The poem attributed to ʿAlī ibn Abī Ṭālib that describes the signs concludes they represent God's supreme name.[49] Al-Būnī provides the same interpretation, and he emphasizes this particular name's great supernatural force.[50] The

43 Late-thirteenth or fourteenth-century manuscript copy of *Shams al-maʿārif*: Arabic text in Winkler 1930, 68; German translation in ibid., 70–71. Printed edition of *Shams al-maʿārif al-kubrā*: Arabic text in al-Būnī, n.d., 1:80–81; German translation in Winkler 1930, 72–74.

44 Among the great many definitions of magic, Pielow's (1995, 26) captures the essence best. See also Savage-Smith 2004, xiii.

45 It cannot be the aim of this chapter to review the concept of *siḥr* in Islam. For this, see Fahd 1997, 567b–571b; and Dorpmüller 2005, 4–7 and 23–38. Fahd also points out that the authors dealing with the topic vary on the specifics. Savage-Smith (2004, xiii–li) provides a very useful overview of the various practices that make up Islamic magic.

46 See the printed edition of *Shams al-maʿārif al-kubrā*: al-Būnī, n.d., 1:81–82 and 85–86; and Winkler 1930, 74–77 and 81–84.

47 Ullmann 1978, 1097b–1098a; and Fahd 1971, 595b–596b. See Fahd 1966, 214–245, for further details.

48 For details, see Dorpmüller 2005, 23–38. It is important to acknowledge that this classification is not a Western projection.

49 Ewald 1839a, 108.

50 See El-Gawhary 1968, 118–165.

group of seven signs is one of the shapes that the name can take, encoded in strange and unfamiliar symbols that do not reveal their true meaning to the uninitiated. When handled appropriately—namely, inscribed ("stamped") on an object—however, they will bestow on the object's user the benefit of the name's various powers. In accordance with that function, the signs have been called the "seven magical signs."[51] Although this kind of magical practice calls upon God, it is based on an understanding of a specific relationship between God and human beings, one in which both sides can affect each other,[52] and it therefore goes beyond true religious devotion. In Yemen, the group of seven signs is found in similar contexts to the first sign, in particular on domestic and religious architecture. For instance, it is inscribed three times on the exterior walls of the mosque of al-Aʿwar, a hamlet in the vicinity of al-Jabīn: over the west window (fig. 121) and on either side of the prayer niche on the north wall.[53]

A second example, the cistern of Bayt al-Shaykh, also carries a marked religious imprint. The group of seven signs is flanked on the right by the *shahāda* (fig. 118). As noted, the cistern adjoins the Grand Mosque of al-Jabīn and partly collects its water from the mosque's roofs; in turn, any overflow feeds the mosque's ablution facilities. God's supreme name (the group of seven signs) was written on its enclosing wall in the hopes that its supernatural force would influence the course of natural events and ensure the rains, indispensable for filling the reservoir. The group of seven signs on the cistern is dated by the building inscription to 1885/6. Clearly, use of the seven signs was not just a medieval phenomenon.[54]

2 Communal Rain-Making Rituals

2.1 *In Pre-Islamic Yemen*

In Yemen, more forceful attempts to urge God to let it rain go back to pre-Islamic times. The inscription Ja 735 (+ Ja 754), dating from the early third century CE, was discovered in 1951–1952 during excavations of the Awām temple in the

51 Porter 1998, 145.

52 See Goldziher 1906, 1:303–308 and 313.

53 Based on the painted ceiling decorations and Kufic inscriptions, the mosque can be dated to around 1200. Interestingly, a threefold magic square is found on the east wall. On the significance of threefold magic squares, see Macdonald 2004, 153a–154a. The north wall also includes a stone with Ancient South Arabian writing. For other details of the building, see al-ʿAṭṭāb 1987, 103–104; and al-ʿArūsī 2003, 865.

54 For other examples of the group of seven signs on domestic architecture in Yemen, see Giunta 2002, figs. 11 and 12; and Wöhrlin 1999, no. 61.

FIGURE 121 Masjid al-Aʿwar: group of seven signs (including six-pointed star) above a window on the west wall

south oasis of Mārib, some 3.5 km away from the city (see fig. 5).[55] The temple was dedicated to the god ʾAlmaqah, the principal Sabaean deity, and hence it had a special status as a centre of pilgrimage. Its large open-air inner courtyard (with a long axis measuring some 100 m) offered space for several thousand people. The text describes the desperate situation caused by the complete failure of the rains in three consecutive seasons. The fields were barren, the trees

55 The first (and today regarded as outdated) translation of the inscription was prepared by Jamme (1962, 211–213 and plate 36). The following is based on Beeston 1972, 352–353; Ryckmans 1973, 379–388; Müller 1986, 10–11; and Müller 1988a, 450–452.

had died, and the wells had dried up. In this state of emergency, the entire tribe living in and around Mārib, as well as the "daughters of Mārib," undertook a procession to the temple, where the religiously inspired rain-making rituals involved invocations and sacrifices to convince ʾAlmaqah to send rain. Unfortunately, the text is not very specific with regard to the ritual act, but it is evident that women, the aforementioned daughters of Mārib, played a significant role in it. Jacques Ryckmans, who wrote a detailed comment on the inscription, points out that these women are explicitly referred to as not having a tribal affiliation; women standing outside the social order of the tribe include sorceresses, magicians, procuresses, and prostitutes.[56] Through their incantations and offerings, the daughters of Mārib were to implore ʾAlmaqah and to persuade him to finally let it rain. As part of the ritual, they shed a liquid, though it is not clear whether this would have been water, their tears, or even their own blood from having gashed themselves.[57] The inscription states that the efforts were successful: a thunderstorm started in the mountains while people were on their way back from the temple. The rains continued for the rest of the day, and the *sayl* arrived in Mārib during the night. The events happened at the beginning of the late-summer rainy season after almost two years of drought. The first lines of the inscription express everybody's gratitude to ʾAlmaqah.

Ja 735 (+ Ja 754) is not the only text from ancient South Arabia that references rain-making rituals. From other examples, we learn that the ceremony was performed as early as the first half of the first millennium BCE and since then at numerous places in times of need.[58]

2.2 *Attitudes and Practices during Early Islamic Times*

Of course, with the arrival of the new religion of Islam in the seventh century, the rains remained as unreliable as they had been in pre-Islamic times, with the same great temporal and spatial variability—after all, the climate did not change. Therefore, the need for rain-making rituals persisted, and it is not surprising that they continued to be performed, both in Yemen and elsewhere. Yet, even though it is well known that "Islam has incorporated numerous features of a pre-Islamic Arabian heritage, common to the paganism of Hijaz as

56 Ryckmans 1973, 381–382.

57 The last is described in the Hebrew Bible (I Kings 18:28) as having been practiced by the priests of Baal during a rain-making ritual; see Beeston 1972, 352.

58 Beeston 1986, 10–11; see also Ryckmans 1966, 491–492.

well as that of South Arabia,"[59] one inevitably wonders how Muslim religious authorities came to terms with the rain-making rituals.

Muḥammad's uncle al-ʿAbbās enjoyed a particularly high reputation for his ability to cause rainfall.[60] Muḥammad himself is reported to have first participated in a procession and prayer for rain when he was a young boy.[61] According to the narratives of the sayings and exemplary way of acting attributed to the prophet, later in his life, Muḥammad was directly or indirectly involved in successful supplications for rain on several occasions, which underlines the ongoing importance of rain-making rituals. As an example, the *Ḥadīth* compendium of al-Bukhārī (d. 870), considered by Sunni Muslims as authentic (*ṣaḥīḥ*), contains an entire section on invoking Allah for rain—namely, Book 17, "Abwāb al-istisqāʾ" ("The chapters concerning *istisqāʾ*"). Here we find the following report:[62]

> I heard Anas bin Mālik saying, "On a Friday a person entered the main Mosque through the gate facing the pulpit while Allāh's Apostle was delivering the Khuṭba.[63] The man stood in front of Allāh's Apostle and said, 'O Allāh's Apostle! The livestock are dying and the roads are cut off;[64] so please pray to Allāh for rain.'" Anas added, "Allāh's Apostle raised both his hands and said, 'O Allāh! Bless us with rain! O Allāh! Bless us with rain! O Allāh! Bless us with rain!'" Anas added, "By Allāh, we could not see any trace of cloud in the sky and there was no building or a house between us and [the mountain of] Silaʿ." Anas added, "A heavy cloud like a shield appeared from behind it (i.e., Silaʿ mountain). When it came in the middle of the sky, it spread and then rained." Anas further said, "By Allāh! We could not see the sun for a week."

In order to understand the procedure, a second *ḥadīth* needs to be cited here:[65]

> The Prophet went towards the Muṣallā[66] and invoked Allāh for rain. He faced the Qibla[67] and wore his cloak inside out, and offered two Rakʿāt.[68]

59 Ryckmans 1983, 14. One example is the official Muslim pilgrimage to Mecca, the *ḥajj*.
60 Goldziher 1890a, 2:108; Goldziher 1906, 1:309; and Nöldeke 1898, 25–27.
61 Ibn Saʿd/Mittwoch, ed. 1905, 1, 1:54.
62 Al-Bukhārī/Khān, trans. 1971, 2:67–68 (Book 17, *bāb* 5, no. 126).
63 *Khuṭba* means "sermon."
64 I.e., there is no transport because the draught animals have starved.
65 Al-Bukhārī/Khān, trans. 1971, 2:66 (Book 17, *bāb* 3, no. 125).
66 *Muṣallā* means "open prayer place, outside a town."
67 Wehr, 1979, s.v. *qibla*: "direction to which Muslims turn in praying (toward the Kaaba)."
68 Ibid., s.v. *rakʿa*: "a bending of the torso from an upright position, followed by two prostrations (in Muslim prayer ritual)."

There are numerous other reports of supplications for rain in the *ḥadīth* literature.[69] From these we learn that a sequence of steps is involved: a communal prayer of two *rakʿas*, performed in the morning in the open air outside the town, is followed by two sermons, for the first of which the cloak is turned inside out; the actual supplication for rain and imploring God's forgiveness complete the ritual. The participants should abstain from luxury and wear ordinary clothes; fasting, almsgiving, and other pious deeds are recommended.[70]

Clearly, in early Islamic times it was understood that, due to the fact that there was a continual and real need for them, the rain-making ceremonies could not be completely abolished. However, a precise ritual was established for the rogational rite for rain (*istisqāʾ*) in order to distance believers from the pagan practices.[71] The ancient rite had to be appropriated to Islam's requirements by eliminating elements with an openly pagan character. In Islam, *istisqāʾ* is, strictly speaking, a prayer, and this is also how religious authorities from Yemen describe it.

2.3 *Practices in Yemen during Islamic Times*

It is not the purpose of this chapter to review the entire Yemeni literature dealing with *istisqāʾ*. Three examples will suffice to illustrate some important points. First, al-Hādī ilā l-Ḥaqq (d. 911), who had arrived in the far north of Yemen towards the end of the ninth century and established the (Shiite) Zaydi imamate there, set up clear rules for the *istisqāʾ* prayer. The jurist, *ḥadīth* scholar, and historian al-Kūfī (d. early 10th c.), who had joined al-Hādī in Yemen from his native Kufa, lists the details of how the prayer (*ṣalāt al-istisqāʾ*) should be performed, citing al-Hādī.[72] The procedure closely follows what has been described above, including the turning inside out of the cloak. Second, for Sunni Zabīd, Ibn al-Daybaʿ reports in the supplement to his history of Zabīd that on Friday, Muḥarram 13, 904 (August 31, 1498), the local judge (*qāḍī*) ordered the people of Zabīd to get ready for an *istisqāʾ* prayer to be performed the following Tuesday outside the city, in the desert. The preparations included almsgiving and three days of fasting.[73] Again, the description focuses on the religious practices. After all, Ibn al-Daybaʿ was a professor of *ḥadīth* in Zabīd's Grand Mosque. He closes his account with the remark that substantial rainfall set in before the ceremony had even started (on Monday).

---

69 See Wensinck (1927) 1960, 201–202, for a detailed list.

70 Bel 1927, 601b–602a.

71 Fahd 1978, 270a. See Guillaume 1955, 701, for idol worship as part of rogational rites for rain in pre-Islamic Mecca.

72 Al-Kufi 1993, 61–62. For brief information on al-Kūfī, see Kaḥḥāla 1957–1961, 10:54.

73 Ibn al-Daybaʿ/Chelhod, ed. 1983, 244.

Third, a more recent example is a letter written in the southern Yemeni city of Sayʾūn, addressed to the sultan and dated 1931. It outlines habits and customs among women that should be rectified and includes the statement "that women are prohibited to go out on the day of *istisqāʾ* [to the public place], even to the houses that are close to the place of *istisqāʾ*."[74] While the letter does not give any details regarding the ceremony, it is clear that participation is only open to men. One may infer that practices other than prayer would have been performed that would not be considered suitable to be seen or heard by women, hence the restrictions on their being in close proximity. But *istisqāʾ* itself is not prohibited.

In the rogational rites for rain that one can observe to this day in Yemen, the participants chant monotonous invocations as they march in a procession to a place outside a town where they perform a communal prayer, typically in a large open-air enclosure. Depending on local custom, the ceremony includes the visitation of a saint's tomb, its (sometimes dance-like) circumambulation, and its anointment. Eventually, an animal that has been part of the procession is slaughtered (a goat, sheep, or even a steer or a camel). The sacrificed animal is either left in the open as an offering to God or it is eaten in a communal meal. Other kinds of offerings are possible. In some regions of Yemen, the attendees are exclusively men (and often boys); in others, women (and often girls) actively participate. In all cases, rogational rites for rain are very much a communal experience.[75]

---

74 Rodionov and Schönig 2011, 116 (Ar. text), 118 (Eng. trans., slightly amended). For a brief summary of the text, see also Boxberger 2002, 140.

75 Descriptions can be found in the following works: Bury 1915, 105–106 (west of Manākha); Gingrich 1993, 154–156 (in the tribal territory of the Munabbih in the far northwest of Yemen, where the rain-making ritual is called *istighātha*); Ho 2006, 200 (outside Tarīm); Rodionov 1997, 108–112 (at Mawlā Maṭar, a place whose name literally means "Patron of Rain," in southern Ḥaḍramawt; see Harrower 2016, 17, for a photograph of the saint's tomb and sanctuary); Serjeant 1964, 75–76 (in western Ḥaḍramawt); Serjeant 1974, 32 (east of Manākha and in the Wādī Ḍahr, where the ceremony is called *tasqiya*); Serjeant 1983b, 314 (outside Ṣanʿāʾ); Varisco 1982, 105–107 (northwest of Ṣanʿāʾ in al-Ahjur, where the ceremony is also referred to as *tasqiya*). In late March 1987, Keall (personal communication, June 2008) observed a rain-making procession on the road from Ṣanʿāʾ to Ḥadda in response to a delay in the onset of the seasonal rains; it had both male and female participants, who were chanting invocations and leading a sacrificial sheep. Caton (2006, 45–50) describes rain prayers in the Grand Mosque in Ṣanʿāʾ, which are therefore a special case—namely, a version authorized by religious functionaries. (Some shortcomings with regard to the explanations and interpretations given may be due to the fact that the rain prayers are not the main focus of the publication but are used as an ethnographic example to illustrate statements of a theoretical nature.) Rodionov (1999, 120) speculates that in western Ḥaḍramawt the irrigation supervisors seem to be actively involved in the

Parallels with the pre-Islamic rain-making ritual are obvious. Serjeant describes rogational rites for rain at a saint's tomb in the southern Yemeni province of Ḥaḍramawt as "a ceremony so essentially pagan that it might almost serve as commentary to certain pre-Islamic inscriptions."[76]

Clearly, the *istisqāʾ* ceremonies in Yemen show that today, as in the past, ordinary people consider prayer alone an inadequate, slimmed-down version; they demand supporting measures to increase the powers of persuasion.[77] This is not surprising. One cannot change or remove parts of a well-established series of actions if one expects a ritual to remain effective. The repetitive, monotonous chanting of invocations and short formulas has the character of a powerful incantation; it stirs enthusiasm among the participants and, together with the animal sacrifice, strongly reinforces the demand for rain. The rituals at the saint's tomb pursue a similar aim. It is firmly believed that the prospect of success is increased by the special prestige of the deceased, his God-given *baraka* (literally, "blessing"), a superhuman power that is beneficent, miraculous, and wonder working.[78] The practices are a forceful attempt to put the saint in a position where he cannot but help the people through intercession with God.[79]

2.4 *Magical or Religious Practice?*

In Yemen (as elsewhere in the Islamic world), the rogational rites for rain include a wide variety of practices that are not directly related to religion. Even the turning inside out of the cloak in the authorized version of the *istisqāʾ* prayer is of questionable religious significance; it can rather be understood as a means of "sympathetic magic." The term goes back to James George Frazer, who explained that magic is based on the hidden supernatural force of a secret

istisqāʾ ceremony. For the communal meal forming part of *istisqāʾ* and its ancient origins, see Sedov and Bâtâyiʿ 1994, 189. The purpose of the ritual hunt in premodern South Arabia, a religiously inspired practice that apparently originates in pre-Islamic times and clearly does not conform to orthodox Islam, has been interpreted as a petition for rain; see Serjeant 1976, 34–39 and 76–77. Appropriate performance of the hunt is required to ensure adequate rainfall. This explanation for the ritual hunt in ancient South Arabia is also given by Beeston (1948, 191–193).

76 Serjeant 1964, 75.

77 See also Goldziher 1890b, 312.

78 This special prestige that saints enjoy in certain parts of Yemen is unfamiliar in others. The reason is a more egalitarian ethic in tribal societies. See Varisco 1982, 107; and Gingrich 1993, 154–155. Varisco's remark that the (Shiite) Zaydis disapprove of the veneration of saints should not be generalized, though; it may apply to some Zaydi communities, possibly in areas with strong tribal traditions, but certainly not to all of them. The Zaydis have their own spiritual leaders, at whose tombs they perform rituals (e.g., in the northern Yemeni town of Ṣaʿda at al-Hādī ilā l-Ḥaqq's tomb).

79 See also Serjeant 1964, 76; additional details can be found in Serjeant 1976, 12–13.

sympathy between things that allows them to affect each other at a distance. Humans can control, even manipulate, this force through appropriate practices. There are two main branches of sympathetic magic. First, according to the Law of Similarity that like produces like, one can produce a desired effect by imitating it. For instance, turning the cloak inside out is done to effect a change in the weather and to turn the barrenness of the land into fertility. Another example (from pre-Islamic Yemen) is that in which the daughters of Mārib shed a liquid during the ancient rain-making ritual. Second, the Law of Contact (or Contagion) implies that once two things have been in contact with each other, they continue to interact and exchange properties even at a distance after the physical contact has ceased.[80]

What is the purpose of magic, on the one hand, and religion, on the other, and is there a clear distinction between the two? These questions have occupied generations of scholars, who have produced an abundance of literature, yet the issue continues to be debated. In the context of this chapter, the section on "The Origins of Religion" from Max Weber's *Economy and Society* is of particular interest. In brief, Weber explains that the original objective of religion was to avert external misfortune in order to gain an economic advantage in everyday life—that is, a straightforward practical concern of this world (without any expectation of a hereafter). Humans can communicate with the supernatural powers in order to get support. For instance, they can ask God through prayer for help in achieving a goal, knowing full well that God may—or may not—hear their prayer. An omnipotent God cannot be compelled by humans to grant their wishes. Conversely, in magical rituals there is a causal link between magical practice and fulfilment of the request: the purpose of magic is to urge God, to coerce a positive outcome. Magic takes a more manipulative approach than religion. This aspect has already been addressed briefly in the broad definition of magic given above in section 1.2. However, Weber points out that there is no clear dividing line between magical and religious ritual: in almost all human societies, religious practices are strongly penetrated by magical practices.[81]

It is, indeed, a common phenomenon of religions that their rituals are rarely "pure" when dogmatic criteria of religious functionaries are used as a yardstick.[82] Regarding the rogational rites for rain as performed in Yemen, prayer is an important component, but people devote themselves to additional

80 Frazer 1911, 52–54. While most of Frazer's work is regarded as outdated, this description of the mechanisms of magical practices continues to be relevant and useful.

81 Weber 2001, 121–157, especially 121–122 and 153–157.

82 Riesebrodt 2008, 29.

persuasive efforts beyond faith and trust in God alone. For those human actions that are performed with reference to religion but go beyond acts of religious devotion because their purpose is fulfilment of a more or less selfish goal, Rudi Paret coined the term "religious magic."[83] It implies that magical and religious practices do not necessarily have to be antipodes, and this is what one can observe in the rogational rites for rain not only in Yemen but elsewhere as well. They have been appropriately described "as an integral whole, which includes Islamic beliefs, their popular interpretation, and non-Islamic local beliefs."[84] The last two are very much influenced by pre-Islamic traditions. This is also why common features outweigh regional variations in the performance of *istisqāʾ* in Yemen. Traditions have proven to be stronger than, for instance, religious differences between Sunnis and (Shiite) Zaydis, at least until recently.

2.5 *Interreligious Encounters in Rain-Making Rituals*

There is, however, a remarkable religious peculiarity when it comes to rain-making rituals in Yemen. In the past, in times of prolonged drought, the members of the Jewish minority were asked by their Muslim neighbours, including even the imams of Ṣanʿāʾ (the Shiite Zaydi rulers of North Yemen prior to the revolution in 1962), to pray for rain. Usually, the ceremony took place in the Jewish cemetery, Torah scrolls included.[85] We can observe here another example of "the other," in this case the Jews, who are believed to possess extraordinary powers that allow them to achieve something that nobody else can.[86] But these powers, much sought after in times of misfortune, had a downside. From time to time, the Jews were, in fact, accused of using them to cause harm—for instance, to prevent rainfall.[87]

There is an additional interesting detail here. In Ṣanʿāʾ, no later than the 1920s, the ceremony had a marked religious character, which was, in fact, a relatively recent development. In the second half of the nineteenth century, the Jewish rain-making rituals did not only involve but were based on the performance of incantations and other magical practices. They were considered

83 Paret 1958, 86.

84 Abu-Zahra 1988, 510. While the author's concern is to show how in the rain rituals, "Islamic thought penetrates popular imagination" (526), one could also reverse the argument: the rain rituals are an example of the popular imagination penetrating Islamic thought.

85 Eraqi-Klorman 2009, 128. See Serjeant 1983b, 314, for a different kind of Jewish-Muslim ceremony in Ṣanʿāʾ.

86 The point has already been made in Case Study 6, section 7, and in Case Study 7, section 4.

87 Eraqi-Klorman 2009, 128.

essential to remove the spell that shut up the heavens; prayer alone was not sufficient.[88]

3 Conclusion

The common principle of the performance of rogational rites for rain (*istisqāʾ*) in Yemen and the use of the group of seven signs as observed on the cistern of Bayt al-Shaykh is an attempt to establish control over nature, in the case of the former through powerful rituals and incantations in times of acute crisis, and in the example of the latter as a preventive measure through invocation of supernatural powers that have their origin in God's supreme name. Both practices are ultimately addressed to God but use means that fall outside the scope of religious devotion. They can best be termed "religious magic" and reflect the feeling of helplessness against the forces of nature in an environment where whether fundamental human needs—most importantly water—are met or not is left to chance.

In al-Jabīn, people had the engineering skills to build cisterns and associated water-harvesting devices. Technical knowledge was not the issue. The same is true of Mārib, where sophisticated technical skills enabled the construction of a massive irrigation scheme. However, in both examples, the rains needed to feed the systems were unreliable, and water scarcity was a perpetual threat to the sustainability of life. Under the circumstances described in this chapter, practices of a magical nature that seek to ensure a guaranteed water supply acquire their own validity.

88 Brauer 1934, 366–367; see 7–11 on his informant.

Concluding Remarks

This book deals with the three-way relationship between water, land, and humans in Yemen's history. Because of the arid or semi-arid climate in large parts of the country, the land itself has little value unless the scarce water resources can be harnessed by human intervention. Since ancient times, humans have made good use of water both for irrigated agriculture and for domestic consumption. As the eight case studies illustrate, the individual ecosystems necessitated different engineering and management approaches.

The oasis of Mārib (Case Study 1) serves as a prominent and typical example of the ancient settlements with their agricultural hinterlands, mainly located along the foot of the highlands. The biannual rainy seasons in the catchment area of the wadis result in short-lived spates in spring and late summer, which were diverted for irrigation farming. Control of the *sayl* with increasingly sophisticated hydraulic structures was decisive for Mārib's economic prosperity and culminated in a diversion barrage built across the mouth of the gorge where the wadi debouches from the mountains. Wells and cisterns, the latter filled by diverting water from the *sayl*, provided a source of water year-round for small-scale irrigation as well as for drinking (by humans and animals) and domestic use.

Sayl irrigation continued to be practiced in Yemen after the rise of Islam, primarily because the seasonal wadi spate is the only source of water for large-scale irrigation in major parts of the country. For instance, in the Wādī Zabīd (Case Study 2), irrigation schemes were put in place from the ninth century onwards. We find the same principle of diverting the *sayl* in a controlled manner in the Wādī Zabīd as in the ancient oasis of Mārib but with the key difference that a diversion barrage was not built across the mouth of the gorge; because the gorge is less distinct, a succession of diversion devices were constructed on either side of the wadi bed. *Sayl* irrigation maintains the fertility of the soil for two reasons. First, the water ensures periodic deposition of nutrient-rich sediments. Second, submerging the fields with large volumes of water results in salts being leached into the subsoil; salinization of the top soil layers is avoided. Furthermore, *sayl* irrigation does not overexploit the water resources: it uses the seasonal spate that flows down the wadi following rainfall in the catchment area. The Wādī Zabīd and the ancient oasis of Mārib are long-term success stories of sustainable irrigation farming; both are human-made landscapes that bear the marked imprint of many centuries of irrigated agriculture.

It was the staple crops from the land in the Wādī Zabīd that allowed the city to prosper and a vibrant religious and scholarly life to develop (Case Study 3).

 | DOI:10.1163/9789004387713_016

An inscription dated 1533 documents a charitable endowment (*waqf*) for the support of a religious college (*madrasa*). The text lists individual properties in the Wādī Zabīd, named by the canals irrigating them, and the harvest levies derived from them. The terminology of the *waqf* deed and many of the names that it specifies are familiar from the historical sources; they are also to a large extent still in use in the traditional *sayl*-irrigation system today. The inscription nicely fits the overall picture of continuity and stability in the Wādī Zabīd.

In addition to the diversion of the *sayl* in the Wādī Zabīd, underground canals and glazed earthenware pipes tapping into the groundwater or the perennial base flow of the wadi were built in cut-and-fill style (Case Study 4). They brought particle-free, clear water to suburban garden estates that required year-round irrigation. After all, passion for horticulture and agriculture was widespread in medieval times, even among the rulers themselves, who were also supportive of introducing new plants to Yemen.

Zabīd's urban water-supply system (Case Study 5) was fed by wells and consisted of holding tanks, open and closed canals, and lead pipes. As medieval Zabīd was a city with numerous religious institutions, it was a priority to supply mosque ablution facilities with water, including public bathing facilities and toilets. The city residents preferred to source drinking water from private wells. Of course, whether in the city or farther up in the wadi, the groundwater that was tapped was directly dependent on the *sayl* to replenish it.

It was also groundwater, albeit in an entirely different environment, that gave rise to the southern Yemeni town of Ghayl Bā Wazīr (Case Study 6). Starting around the fourteenth century, subterranean water in a gypsum karst was tapped by gently sloping underground canals and brought to the surface under gravity flow several kilometres to the south of the gypsum shield, where soil for plant cultivation is found. These canals are locally called *maʿāyīn*, sg. *maʿyān*. The engineers worked with precision in order to maintain a low gradient and avoid depletion of the subterranean water resources that would have rendered the system unsustainable. Vertical shafts were dug at intervals to provide access to the underground canal and allow removal of the excavated debris. Unlike the seasonal wadi spate, the *maʿyān* system provided a dependable supply of water year-round, an invaluable resource in an otherwise bleak and arid environment. Since the water was ultimately clean groundwater, it was used first for domestic purposes by the inhabitants of the settlement that developed on the southern fringes of the gypsum shield. The grey water was then directed to the vegetable gardens and orchards beyond the town. Because of the high level of calcium in the water from the karst, drinking water came mostly from a *maʿyān* drawing sweet water from gravel beds on the far side of the gypsum shield. Westerners usually associate underground canals that have vertical shafts set at regular intervals along their course with the famous *qanāt* systems of Iran and credit Iranian engineers with having introduced the technical principle to Yemen. However, it is argued here that, instead of there having been a technology transfer, the engineering approach taken to exploit the subterranean water was developed locally in response to Ghayl Bā Wazīr's hydrogeological and geomorphological conditions. There is no need to seek a precedent outside of Yemen for the technical traits observed.

The way in which local characteristics define water-engineering principles can also be observed in al-Jabīn (Case Study 7), a town perched on the western escarpment of the mountains of Yemen, where arable farming is limited to man-made terraces. Those on the western slopes facing the Red Sea receive precipitation during the two rainy seasons. The east-facing terraces lie in the rain shadow and require rainwater harvesting from the adjacent slopes. Because of al-Jabīn's geographical position, digging wells is not an option, and perennial springs supplying the residents with drinking water are only found at much lower levels down the mountains. Nonpotable water for domestic consumption comes from open cisterns. Like *sayl* irrigation, rainwater harvesting makes use of renewable water resources—namely, the surface runoff from the mountain slopes. Unlike the *sayl*, which, to a large extent, is the result of natural flow patterns, the runoff rainwater is harvested from cleared or plastered surfaces through man-made diversions; it is (mostly) the result of human intervention. Rainwater harvesting entails a built environment.

There is an interesting linguistic detail here: the Arabic term *birka* describes both a cistern and also the catchment area with its rainwater-harvesting structures, on which the collection of surface runoff depends; the storage reservoir and the catchment area feeding it constitute a unit. A different kind of example illustrating the conception of land and water as one integrated whole can be found in the terminology of *sayl* irrigation. For instance, the term "wadi" refers to the flood course as well as to the land watered by it; a *sharīj* is a primary canal and also the contiguous fields that it irrigates; *maʿqam* designates both the small temporary dike across a subordinate canal and the fields that receive the water from it.

It is not surprising that water-related issues form a recurring theme in local folklore and storytelling. Legends revolving around water often have some factual basis and refer, for instance, to past accomplishments in hydraulic engineering and the natural environment in which they evolved. On occasion, when the engineering schemes seem too grand and therefore impossible to achieve under current circumstances, local folklore credits their construction to "the other," non-Arab strangers in Yemen (for instance, the Ottomans) or non-Muslims (the ancient South Arabians or members of the Jewish minority).

They are suspected to possess the extraordinary, perhaps supernatural, powers required for such feats of engineering. Examples are the *maʿyān* system of Ghayl Bā Wazīr and the cistern of al-Jabīn.

While engineering skills were needed to build the infrastructure, it was equally important to keep it in good repair and to devise strategies to allocate the water equitably. That is, a water-engineering scheme is not only a technical issue but also a social system of negotiation, arbitration, and mutual agreement. Sharing water requires regulations, and sharing scarce water requires even more stringent but fair regulations in order to avoid constant clashes. This principle is the same on all four sites studied, from ancient Mārib to premodern al-Jabīn. The water-allocation rights integrated customary rules of conduct that were shaped in response both to the natural environment and to the community's needs. Hence their details differ from one location to another. Since ancient times, water allocation, maintenance, and repair work had been supervised by one chosen person. While problems did occur in day-to-day operations, they were manageable. Smooth operation of the system was to everybody's advantage, both individual and collective, and required a spirit of general consensus on key issues, communal responsibility, and cooperation. Together with consideration for the natural resources, these are prerequisites for the longevity of a water management system, as documented in Mārib, Zabīd, Ghayl Bā Wazīr, and al-Jabīn.

Yet one fundamental problem common to all the sites studied was beyond human control. The rains on which the operation of the engineering schemes ultimately depended were unreliable. Successful farming and the very survival of the communities were jeopardized by severe droughts that occurred at irregular intervals. Engineering or managerial skills could not provide the solution. Therefore, people resorted to an entirely different kind of approach to try to establish control over nature (Case Study 8). In times of severe crisis, they performed powerful communal rituals to urge God to let it rain. This practice has its origins in pre-Islamic times, is documented in written sources from the Middle Ages, and can be observed to this day. As a preventive measure, a group of seven signs symbolizing God's supreme name was inscribed on, for instance, a cistern in the hopes that its supernatural force would bring about the rains to fill the reservoir. Both practices can be designated "religious magic," and they show people's helplessness in the face of unreliable or completely absent rainfall.

It may sound like an irony that it was not scarcity of water but rather an overabundance of it that brought an end to the ancient oasis of Mārib: a disastrous out-of-control flood in the late sixth century CE fatally damaged the diversion barrage and led to the abandonment of the site. Even though there was an underlying technical problem—namely, the loss of stability of the hydraulic structures caused by raising them again and again to compensate for the rising field level and the reduction in gravity flow—the example shows the devastating effect of too much water. In this context, it is interesting to note that since the arrival of Islam, rain rogations have been performed in situations of acute crisis, including not only prolonged draught but also excessive flooding.[1] Both extremes, too little as well as too much rain, are equally destructive and beyond engineering and managerial skills to control.

---

1 For instance, in the first *ḥadīth* cited in Case Study 8, section 2.2, after it is stated that the sun was not seen for an entire week because of heavy rainfall, the report continues:

"Anas further said: "... Next Friday a person entered through the same gate and at that time Allāh's Apostle was delivering the Friday's Khuṭba. The man stood in front of him and said, 'O Allāh's Apostle! The livestock are dead and the roads cut off, please pray to Allāh to with-hold rain.'" Anas added: "Allāh's Apostle raised both his hands and said: 'O Allāh! Round about us and not on us. O Allāh! On the plateaus, on the mountains, on the hills, in the valleys and on the places where trees grow.' So the rain stopped and we came out walking in the sun." (Al-Bukhārī/Khān, trans. 1971, 2:68 [Book 17, *bāb* 5, no. 126])."

The *ḥadīth* also emphasizes once more an understanding of the cultivated land as constituting a unit with its catchment area, where the rains that are essential for successful irrigation farming fall.

Glossary

Arabic terms that are common in English (e.g., "imam" or "wadi") are neither transliterated nor italicized; their correct Arabic transliteration is given in parentheses.

ab "father well": large well feeding a *maʿyān*; see also *umm*

al-aʿlā fa-l-aʿlā "the one who is highest [with his plot], then the one who is next highest": sequence of water allocation by upstream priority according to *sharīʿa*

alif first letter of the Arabic alphabet

ʿāqil*, pl. *ʿuqqāl neighbourhood headman

ʿaqm*, pl. *ʿuqūm 1) diversion barrage erected across a wadi; alternative term: *maʿqam* (sense 3)
2) diversion barrage that diverts part of the *sayl* into a primary canal (*sharīj*); alternative term: *maʿqam* (sense 4)

ʿayn outlet of a water reservoir (*jābiya*)

baraka blessing

basmala formula that constitutes the first line of all the chapters of the Koran except the ninth: "In the name of God, the Merciful, the Compassionate"; invocation of the *basmala* should precede any important act in a Muslim's life, calling for its divine blessing and consecrating it

baṭṭ deep open section of a *maʿyān*, sometimes with bridging arches to provide stability and prevent wall collapse, but otherwise unroofed

birka*, pl. *birak 1) open ablution pool in a mosque complex
2) open cistern that collects surface runoff following a rain, based on the principles of rainwater harvesting
3) open cistern and the entire catchment area with its rainwater-harvesting structures on which the collection of runoff in the cistern depends

dirham (*dirham*) 1) unit of weight
2) unit of silver currency

***fajr* prayer** dawn prayer; one of the five daily prayers

fiqh jurisprudence

ghayl*, pl. *ghuyūl 1) general meaning: perennial stream
2) in the context of a wadi: base flow
3) in colloquial Yemeni usage, in an urban context: artificial often partly subterranean canal

ghusl major ritual ablution, involving a thorough washing of every part of the body including the hair

hāʾ letter of the Arabic alphabet

 | DOI:10.1163/9789004387713_017

ḥadath state of minor ritual impurity, stemming from, for instance, contact with a substance that is regarded as unclean, such as wine, dogs, or excrement

ḥadīth*, pl. *aḥādīth "report": narrative communicating a *sunna* of the Prophet Muḥammad

Ḥadīth corpus of *aḥādīth*

ḥammām*, pl. *ḥammāmāt ablution facility for a mosque, providing a space for private bathing

ḥarra*, pl. *ḥarrāt round opening set into a *marda*ʿ, usually with a diameter of 15–20 cm, that can be blocked and unblocked to regulate the flow in a *maʿyān*

ḥashāsh volcanic cinder (can be used as aggregate in *qaḍāḍ*)

ḥawḍ*, pl. *aḥwāḍ water trough; holding tank for water

ḥawma*, pl. *ḥuwam collapse sink or collapse doline; it is filled with water year-round

hilsin grey grit (used as aggregate in *qaḍāḍ*)

huwa "He": Koranic designation for Allah

imam (*imām*) 1) in Sunni Islam: leader in the congregational prayer
2) in Shiite Islam: divinely inspired political and spiritual leader of the community who is a descendent of the Prophet Muḥammad through his daughter Fāṭima

***istisqā*ʾ** rogational rite for rain

jābiya*, pl. *jawābī water reservoir

jalla*, pl. *jalal field created through land reclamation in the low-lying areas along the sides of the wadi bed; it is irrigated by a small barrage (*manṣūb* or *muḥammal*)

***jāmi*ʿ** grand mosque; central mosque in an Islamic city where the communal prayer is performed on Fridays

janāba state of major ritual impurity that ensues from sexual intercourse as well as from menstruation and childbirth

jinn spirits, demons

jirba*, pl. *jirab field that is fed by occasional rainfall (not irrigated from the *sayl*)

jisr*, pl. *jusūr spillway; engineered device built between fields or into the course of a canal to accommodate a drop in elevation between terraces

kahra water vapour forming around natural joints and fissures (see *khawʿa*) in a karst, indicating underground water

kanīf*, pl. *kunuf public toilet

karīf rain-fed and therefore periodically dry waterhole

karrī niṣf coarse gravel, measuring up to approximately 3.5 × 2 × 1 cm (used as aggregate in *qaḍāḍ*)

***karrī rub*ʿ** smaller-sized gravel, measuring some 2 × 1 × 1 cm (used as aggregate in *qaḍāḍ*)

kharāb breach in a field bank (*zabīr*) to allow field-to-field irrigation

khāṣṣa*, pl. *khawāṣṣ special properties—that is, unaccountable, hidden powers—of such things as Koranic verses and God's names, including their constituents, most importantly the letters of the alphabet; these secret powers are used in white magic

khātam seal

khawʿa*, pl. *khuwaʿ natural joint or fissure in a karst that percolates water

khudra piece of cloth attached to a rope used to plug the outlet of a reservoir (*ʿayn*) and—by pulling the rope—to unplug it

khuṭba sermon

lajna advisory committee assisting the supervisor of a *maʿyān* (*muqaddam al-maʿyān*)

māʾ kādhiba "lying water" or "false water": groundwater with an unreliable flow

maʿād*, pl. *maʿāwud measure of surface area

madrasa*, pl. *madāris religious college, usually associated with a mosque

***maghrib* prayer** sunset prayer; one of the five daily prayers

manfadh modern version of a spillway (see *jisr*), built with concrete blocks, Portland cement, and plastic pipes

mann*, pl. *amnān measure of capacity

manṣūb*, pl. *manāṣīb barrage consisting of heaped-up earth and coarse gravel and often reinforced with brushwood, palm fronds, and boulders; the smaller kind is found in the low-lying areas along the sides of the wadi bed and diverts part of a *sayl* onto fields created through land reclamation (see *jalla*); the larger size serves as a diversion barrage that diverts part of the *sayl* into a primary canal (*sharīj*); alternative terms: *muḥammal*, *ʿaqm* (sense 2), *maʿqam* (sense 4)

manzil* (or *manzilat*) *al-qamar*, pl. *manāzil "lunar mansion": conspicuous star or constellation that is identified along the moon's path against the background of the fixed stars; there are twenty-eight lunar mansions, of which thirteen or fourteen are visible during any given night of the year, in accordance with the season

maʿqam*, pl. *maʿāqim 1) small temporary dike of mud and brushwood built across an irrigation canal to direct the water into an adjoining field

2) unit of contiguous fields that is irrigated by such a dike; subdivision of a *sharīj* (see *sharīj* sense 2)

3) diversion barrage erected across a wadi; alternative term: *ʿaqm* (sense 1)

4) diversion barrage that diverts part of the *sayl* into a primary canal (*sharīj*); alternative term: *ʿaqm* (sense 2)

mardaʿ*, pl. *marādiʿ stone barrier in the last third of the underground section of a *maʿyān*, usually with one to three round openings (see *ḥarra*) that can be blocked and unblocked individually; a *mardaʿ* acts as a sluice gate

mawkhasha flat, hand-sized pounding stone used to compact *qaḍāḍ* and eliminate voids

maʿyān, pl. ***maʿāyīn*** 1) canal that consists of an underground section (see *shaṭṭ*) and an above-ground section (an open surface canal); the underground section is typically cut through rock, with vertical shafts (see *naqba*) set at regular intervals along its course

2) land that is irrigated by such a canal

miḥrāb prayer niche, recess in a mosque wall indicating the direction of prayer (towards the Kaaba in Mecca)

mīm letter of the Arabic alphabet with the numerical value of forty

muʿallim master builder

mubāḥ something that is not owned, ownerless because it cannot be owned; therefore everybody has the right of free access to it (e.g., water in its natural state)

mudd, pl. ***amdād*** measure of capacity

muezzin (*muʾadhdhin*) person who recites the call to the five daily prayers

mughtasala, pl. ***-āt*** partitioned bathing space for private ablution, at the side of a *birka*

muḥammal see *manṣūb*

muqaddam al-maʿyān supervisor of a *maʿyān*

muṣallā open prayer place, outside a town

nāʿim oblong riverine (and therefore rounded) pebble, used for smoothing and polishing a *qaḍāḍ* surface

naqba, pl. ***nuqab*** vertical shaft that provides access to the underground section of a *maʿyān* in addition to ventilation and light, and facilitates removal of excavated rock

nashaf conical pumice stone for sanding down rough protrusions from a *qaḍāḍ* surface

nūra 1) quicklime (CaO)

2) viscous paste of slaked lime ($Ca(OH)_2$) and water used as a binder in *qaḍāḍ*

qaḍāḍ traditional Yemeni plastering material that is applied in multiple layers to a building's external and internal facades as waterproof lining; the two essential components are a binder (*nūra*) and aggregate; in volcanic regions of Yemen, volcanic cinder (*ḥashāsh*) is used as aggregate

qāḍī judge

qanāt, pl. ***qanawāt*** like a *maʿyān*, with the difference that it is typically dug through alluvial sediments at the foot of a mountain range to tap the aquifer; *qanāt* is also the umbrella term used in the literature for such (partly) subterranean canals that are found with slight technical variations throughout the Middle East and North Africa and called, for instance, *falaj* (in Oman, the United Arab Emirates and Saudi Arabia), *fujjāra/fuqqāra* (foggara) (in North Africa, especially Libya and Algeria), *kārīz* (in eastern Iran and Afhganistan), and *khaṭṭāra* (in Morocco)

qibla direction to which Muslims turn in praying (towards the Kaaba in Mecca)

qīrāṭ*, pl. *qarārīṭ measure of capacity

rakʿa bending of the torso from an upright position, followed by two prostrations (in Muslim prayer ritual)

raṭl*, pl. *arṭāl measure of capacity

ṣaḥīḥ correct, authentic

ṣalāt prayer

ṣaqīʿ*, pl. *ṣuqūʿ small temporary dike to block a canal; it typically consists of stones, mud, grass sods, and cloth

sāqiya*, pl. *sawāqī 1) (smaller) secondary canal
2) small unlined canal-like structure on a hillside for rainwater harvesting, created either by digging a ditch (where possible) or, more commonly, by heaping up gravel and stones; as soon as it rains, the surface runoff from the hillside is funnelled through it onto the lower-lying fields

sawqa deepening of a *maʿyān* along its entire length to maintain a continuous flow in places where the water table has dropped dramatically

sayl*, pl. *suyūl seasonal spate in a wadi

shāfiʿī Shafiitic or Shafiite: adhering to—or an adherent of—the *shāfiʿī* school of law in Sunni Islam, named after its founder Muḥammad ibn Idrīs al-Shāfiʿī (d. 819); the *shāfiʿī* school of law is authoritative for determining water law in large parts of Yemen

shahāda Muslim creed of the oneness of God and acceptance of Muḥammad as God's prophet

sharīʿa codified Islamic law; literal meaning: "place of descent to water" or "way to water"

sharīj*, pl. *shuruj 1) primary canal that is connected to a diversion barrage at its head and that leads towards the fields lying above the banks of the wadi
2) unit of contiguous fields irrigated by such a primary canal

shaṭṭ*, pl. *shuṭūṭ underground section of a *maʿyān*: gently sloping gallery connecting the bottoms of the *maʿyān* shafts (see *naqba*), typically cut through rock

shaykh*, pl. *shuyūkh 1) tribal leader
2) community leader (in non-tribal parts of Yemen, such as al-Jabīn)
3) title that reflects a family's hereditary nobility

shaykh al-sharīj supervisor of an agricultural area irrigated by a specific *sharīj* in the Wādī Zabīd; he bears responsibility for enforcing the water law, for ensuring that repairs and maintenance work are carried out, and for settling disputes among the farmers of the *sharīj* or with other areas

siḥr magic: presupposition of hidden supernatural forces that can be controlled by humans to their advantage through appropriate practices; magic includes white magic (see *sīmiyāʾ*) and black magic (which seeks to harm others and is prohibited)

sīmiyā' white magic: causes no harm to others but rather tries to ward off hardship; permitted in certain cases; see also *khāṣṣa*

sunna*, pl. *sunan saying or exemplary way of acting attributed to the Prophet Muḥammad

Sunna whole corpus of the sayings and exemplary ways of acting attributed to the Prophet Muḥammad

ṭahāra state of ritual purity

taqdūma feeder canal that connects an additional water source (or reconnects the original water source) to an existing *ma'yān* suffering from reduced flow

thumn*, pl. *athmān measure of capacity; literal meaning: "eighth"

ṭīn mud mortar

umm "mother well": large well feeding a *qanāt*; see also *ab*

'urf customary law

usṭā*, pl. *asāṭiya master builder; colloquialism derived from *ustādh*

ustādh*, pl. *asātidha teacher or professor, also used as a title

wadi (*wādī*) 1) general meaning: dry stream bed that contains water only from rain that falls on higher ground in its catchment basin; typical watercourse of arid regions

2) in *sayl*-irrigated areas: land watered by the wadi

waqf*, pl. *awqāf endowment

waqf khayrī public-good *waqf*: charitable endowment for the support of public municipal and religious institutions

waqfiyya *waqf* deed

wāw letter of the Arabic alphabet

wuḍū' minor ritual ablution before prayer—namely, washing the face, hands and feet

zabadī*, pl. *azbūd measure of capacity

zabīr*, pl. *zubur field bank, up to 1 m high, to allow irrigation by deep ponding

zahab*, pl. *zāhīb field that is contained by high field banks (*zubur*, sg. *zabīr*) to allow *sayl* irrigation by deep ponding

Works Cited

Abū Dāwūd, Sulaymān ibn al-Ashʿath al-Sijistānī/ʿAbd al-Ḥamīd, Muḥammad Muḥyī al-Dīn, ed. 1950. *Sunan*. 4 vols. 2nd ed. [Cairo]: Maṭbaʿat al-saʿāda.

Abu-Zahra, Nadia. 1988. "The Rain Rituals as Rites of Spiritual Passage." *International Journal of Middle East Studies* 20:507–529.

Āghā Buzurg al-Ṭihrānī, Muḥammad Muḥsin. 1938/9 [1357 AH]. *Al-Dharīʿa ilā taṣānīf al-shīʿa*. Vol. 3. Al-Najaf: al-Gharī.

Ahlwardt, Wilhelm. 1893. *Verzeichnis der arabischen Handschriften der Königlichen Bibliothek zu Berlin*. Vol. 5. Berlin: Asher.

Al-Hubaishi, Ahmed, and Klaus Müller-Hohenstein. 1984. *An Introduction to the Vegetation of Yemen: Ecological Basis, Floristic Composition, Human Influence*. Eschborn: Deutsche Gesellschaft für Technische Zusammenarbeit (GTZ).

Al-Radi, Selma. 1994. "Quḍāḍ: The Traditional Yemeni Plaster." *Yemen Update* 34:6–13.

Al-Radi, Selma. 1997. *The ʿAmiriya in Radaʿ: The History and Restoration of a Sixteenth-Century Madrasa in the Yemen*. (Oxford Studies in Islamic Art, 13). Oxford: Oxford University Press.

Al Tikriti, Walid Y. 2002. "The South-East Arabian Origin of the Falaj System." *Proceedings of the Seminar for Arabian Studies* 32:117–138.

Al Tikriti, Walid Y. 2011. *Archaeology of the Falaj: A Field Study of the Ancient Irrigation Systems of the United Arab Emirates*. Abu Dhabi: Department of Historic Environment, Abu Dhabi Authority for Culture and Heritage (ADACH).

Anawati, Georges C. 1967. "Le nom suprême de Dieu." In *Atti del terzo congresso di studi arabi e islamici, Ravello, 1–6 settembre 1966*, 7–58. Napoli: Istituto universitario orientale.

Arberry, Arthur J., trans. 1998. *The Koran Interpreted*. (Oxford World's Classics). Oxford: Oxford University Press.

van Arendonk, Cornelis. 1968. "Ibn al-Daybaʿ." In *The Encyclopaedia of Islam*, 2nd ed., III:746a–b. Leiden: Brill.

Arnaldez, Roger, and Louis Massignon. 1963. "Arabic Science." In *Ancient and Medieval Science from Prehistory to AD 1450*, ed. René Taton, 385–421. London: Thames and Hudson.

al-ʿArūsī, Muḥammad ʿA. 2003. "Al-Jabī." In *al-Mawsūʿa al-yamaniyya*, 2nd ed., 2:864–867. Ṣanʿāʾ: Muʾassasat al-ʿafīf al-thaqāfiyya.

al-ʿAṭṭāb, Aḥmad L. 1987. "Mawāqiʿ athariyya jadīda fī qaḍāʾ Rayma." *Al-Iklīl* 5 (2): 94–104.

Ayalon, David. 1975. "Names, Titles and 'Nisbas' of the Mamlūks." *Israel Oriental Studies* 5:189–232.

Aziz, Muhammad Ali. 2011. *Religion and Mysticism in Early Islam: Theology and Sufism in Yemen*. London: Tauris.

Bā Qaḥayzal, Sālim ʿA., Ibrahīm A. Saʿīd, and Muḥammad S. Ibn Ghūth. 1996. *Dirāsa tawthīqiyya li-namādhij min anẓimat al-rayy al-taqlīdiyya wa-ṭuruq al-ḥaṣād al-māʾī fī muḥāfaẓatay Ḥaḍramawt wa-Shabwa*. Ṣanʿāʾ and Aden: Environment Protection Council; al-Mukallā: College of Education (unpublished report).

Baack, Lawrence J. 2014. *Undying Curiosity: Carsten Niebuhr and The Royal Danish Expedition to Arabia (1761–1767)*. Stuttgart: Steiner.

Bagg, Ariel M. 2000. *Assyrische Wasserbauten: Landwirtschaftliche Wasserbauten im Kernland Assyriens zwischen der 2. Hälfte des 2. und der 1. Hälfte des 1. Jahrtausends v. Chr.* (Baghdader Forschungen, 24). Mainz: Zabern.

Balescu, Sanda, Jean-François Breton, Brigitte Coque-Delhuille, and Michel Lamothe. 1998. "La datation par luminescence des limons de crue: Une nouvelle approche de l'étude chronologique des périmètres d'irrigation antique du Sud-Yémen." *Comptes rendus de l'Académie des sciences, Paris, Sciences de la terre et des planètes* 327:31–37.

Beadnell, H. J. Llewellyn. 1909. *An Egyptian Oasis: An Account of the Oasis of Kharga in the Libyan Desert, with Special Reference to its History, Physical Geography, and Water-Supply*. London: Murray.

Beaumont, Peter. 1989. "The Qanat: A Means of Water Provision from Groundwater Sources." In *Qanat, Kariz and Khattara: Traditional Water Systems in the Middle East and North Africa*, eds. Peter Beaumont, Michael Bonine, and Keith McLachlan, 13–31. London: School of Oriental and African Studies.

Beeston, Alfred F. L. 1948. "The Ritual Hunt: A Study in Old South Arabian Religious Practice." *Le Muséon* 61:183–196.

Beeston, Alfred F. L. 1972. Review of *Sabaean Inscriptions from Maḥram Bilqîs*, by A. Jamme. *Bulletin of the School of Oriental and African Studies* 35:349–353.

Beeston, Alfred F. L. 1986. "The Qatabanic Text VL 1." *Proceedings of the Seminar for Arabian Studies* 16:7–11.

Beeston, Alfred F. L. 1995. "Sabaʾ." In *The Encyclopaedia of Islam*, 2nd ed., VIII:663a–665b. Leiden: Brill.

Bel, Alfred. 1927. "Istisḳāʾ." In *The Encyclopaedia of Islam*, 2nd ed., II:601b–602a. Leiden: Brill.

Bent, J. Theodore. 1894. "Expedition to the Hadramut." *The Geographical Journal* 4:315–333.

Bent, J. Theodore, and Maybel V. A. Bent. 1900. *Southern Arabia*. London: Smith.

van den Berg, Lodewijk W. C. 1886. *Le Ḥadhramout et les colonies arabes dans l'archipel indien*. Batavia: Imprimerie du gouvernement.

Berkes, Fikret, Johan Colding, and Carl Folke. 2000. "Rediscovery of Traditional Ecological Knowledge as Adaptive Management." *Ecological Applications* 10 (5): 1251–1262.

Bidwell, Robin. 1976. *Travellers in Arabia*. London: Hamlyn.

Björkman, Walther. 1928. *Beiträge zur Geschichte der Staatskanzlei im islamischen Ägypten*. Hamburg: Friederichsen, De Gruyter.

Blackburn, J. Richard. 1971. "Turkish-Yemenite Political Relations, 1538–1568." PhD diss., University of Toronto (unpublished).

Blackburn, J. Richard 1993. "Muṣṭafā Pasha al-Nashshār." In *The Encyclopaedia of Islam*, 2nd ed., VII:721a–b. Leiden: Brill.

Blackburn, J. Richard. 2005. *Journey to the Sublime Porte: The Arabic Memoir of a Sharifian Agent's Diplomatic Mission to the Ottoman Imperial Court in the Era of Suleyman the Magnificent*. (Beiruter Texte und Studien, 109). Beirut: Orient-Institut; Würzburg: Ergon.

Bliss, Frederick J., and R. A. Stewart Macalister. 1902. *Excavations in Palestine during the Years 1898–1900*. London: Palestine Exploration Fund.

Bloom, Arthur L. 1978. *Geomorphology*. Englewood Cliffs, NJ: Prentice-Hall.

Bonnenfant, Guillemette, and Paul Bonnenfant. 1987. *L'art du bois à Sanaa: Architecture domestique*. Aix-en-Provence: Édisud.

Bonnenfant, Paul, ed. 1995, *Sanaa: Architecture domestique et société*. Paris: CNRS.

Bonnenfant, Paul. 1997. "La maîtrise de l'eau dans le wadi Zabid, Yémen." *Bulletin de l'Association de géographes français* 74:12–24.

Bonnenfant, Paul. 2004. *Zabid au Yémen: Archéologie du vivant*. Aix-en-Provence: Édisud.

Bonnenfant, Paul, and Guillemette Bonnenfant-Outrebon. 1995. "Le qaḍāḍ." In *Sanaa: Architecture domestique et société*, ed. Paul Bonnenfant, 417–445. Paris: CNRS.

Bosworth, Clifford E. 1978. "al-Khazradjī." In *The Encyclopaedia of Islam*, 2nd ed., IV:1188b. Leiden: Brill.

Boucharlat, Rémy. 2003. "Iron Age Water-Draining Galleries and the Iranian '*Qanāt*'." In *Archaeology of the United Arab Emirates: Proceedings of the First International Conference on the Archaeology of the U.A.E.*, eds. Daniel T. Potts, Hasan Al Naboodah, and Peter Hellyer, 162–172. London: Trident.

Bowen, Richard LeBaron. 1958. "Irrigation in Ancient Qatabân (Beiḥân)." In *Archaeological Discoveries in South Arabia*, eds. Richard LeBaron Bowen and Frank P. Albright, 43–131. (Publications of the American Foundation for the Study of Man, 2). Baltimore: Johns Hopkins Press.

Boxberger, Linda. 2002. *On the Edge of Empire: Hadhramawt, Emigration, and the Indian Ocean, 1880s–1930s*. Albany, NY: State University of New York Press.

Boyer, Carl B., revised by Uta C. Merzbach. 1989. *A History of Mathematics*. 2nd ed. New York: Wiley.

Brauer, Erich. 1934. *Ethnologie der jemenitischen Juden*. (Kulturgeschichtliche Bibliothek, 1. Reihe: Ethnologische Bibliothek, 7). Heidelberg: Winter.

Brunner, Ueli. 1983. *Die Erforschung der antiken Oase von Mārib mit Hilfe geomorphologischer Untersuchungsmethoden*. (Archäologische Berichte aus dem Yemen, 2). Mainz: Zabern.

Brunner, Ueli. 1999. *Jemen: Vom Weihrauch zum Erdöl*. Wien: Böhlau.

Brunner, Ueli. 2000. "The Great Dam and the Sabaean Oasis of Ma'rib." *Irrigation and Drainage Systems* 14:167–182.

Brunner, Ueli. 2005. "The Geography and Economy of the Sabaean Homeland." *Archäologische Berichte aus dem Yemen* 10:1–7.

Brunner, Ueli, and Harold Haefner. 1990. "Altsüdarabische Bewässerungsoasen." *Die Erde* 121:135–152.

Brutsaert, Wilfried. 2005. *Hydrology: An Introduction*. Cambridge: Cambridge University Press.

Buffa, Vittoria, and Burkhard Vogt. 2001. "Sabir – Cultural Identity between Saba and Africa." In *Migration und Kulturtransfer: Der Wandel vorder- und zentralasiatischer Kulturen im Umbruch vom 2. zum 1. vorchristlichen Jahrtausend*, eds. Ricardo Eichmann and Hermann Parzinger, 437–450. (Kolloquien zur Vor- und Frühgeschichte, 6). Bonn: Habelt.

al-Bukhārī, Muḥammad ibn Ismāʿīl/Khān, Muḥammad Muḥsin, trans. 1971 [1391 AH]. *The Translation of the Meanings of* Ṣaḥīḥ Al-Bukhārī: *Arabic-English*. Vol. 2. Medina: al-Maktaba al-salafiyya.

al-Būnī, Aḥmad ibn ʿAlī. n.d. [ca. 1905]. *Kitāb Shams al-maʿārif al-kubrā wa-laṭāʾif al-ʿawārif*. 4 vols. Cairo: al-Ḥusayniyya.

Burrowes, Robert D. 2002. "Al-Yaman. 3. History. (b) From 1962 to the Present Day." In *The Encyclopaedia of Islam*, 2nd ed., XI:274b–276a. Leiden: Brill.

Bury, G. Wyman. 1915. *Arabia Infelix or the Turks in Yamen*. London: MacMillan.

Calvet, Yves, and Christian Robin. 1997. *Arabie heureuse, Arabie déserte: Les antiquités arabiques du Musée du Louvre*. Paris: Éditions de la Réunion des musées nationaux.

Canova, Giovanni. 1995. "La *ṭāsat al-ism*: Note su alcune coppe magiche yemenite." *Quaderni di studi arabi* 13:73–92.

Caponera, Dante A. 1973–1978. *Water Laws in Moslem Countries*. 2 vols. (FAO Irrigation and Drainage Paper, 20). Rome: FAO.

Caponera, Dante A. 2001. "Ownership and Transfer of Water and Land in Islam." In *Water Management in Islam*, eds. Naser I. Faruqui, Asit K. Biswas, and Murad J. Bino, 94–102. Tokyo: United Nations University Press; Ottawa: International Development Research Centre.

Carra de Vaux, Bernard, revised by Louis Gardet. 1960. "Basmala." In *The Encyclopaedia of Islam*, 2nd ed., I:1084a–1085a. Leiden: Brill.

Caton, Steven C. 2006. "What Is an 'Authorizing Discourse'?" In *Powers of the Secular Modern: Talal Asad and His Interlocutors*, eds. David Scott and Charles Hirschkind, 31–56. Stanford, CA: Stanford University Press.

Caton, Steven C. 2013. "History." In *Yemen*, ed. Steven C. Caton, 29–87. (Middle East in Focus). Santa Barbara, CA: ABC-CLIO.

Caton Thompson, Gertrude. 1944. *The Tombs and Moon Temple of Hureida (Hadhramaut)*. (Reports of the Research Committee of the Society of Antiquaries of London, 13). Oxford: Oxford University Press; London: The Society of Antiquaries.

Chelhod, Joseph. 1978. "Introduction à l'histoire sociale et urbaine de Zabīd." *Arabica* 25:48–88.

Chorley, Richard J., Stanley A. Schumm, and David E. Sugden. 1985. *Geomorphology*. London: Methuen.

Christiansen-Weniger, Friedrich. 1961. "Alte Methoden der Wassergewinnung für Bewässerungszwecke im Nahen und Mittleren Osten unter besonderer Berücksichtigung der Kanate." *Wasser und Nahrung* 1–2:28–31 and 73–84.

Christiansen-Weniger, Friedrich. 1971. "Bodenbearbeitung." In *Handbuch der Landwirtschaft und Ernährung in den Entwicklungsländern*, eds. Peter von Blanckenburg and Hans-Diedrich Cremer, 2:126–142. Stuttgart: Ulmer.

Ciuk, Christopher, and Edward J. Keall. 1996. *Zabid Project Pottery Manual 1995: Pre-Islamic and Islamic Ceramics from the Zabid Area, North Yemen*. (BAR International Series, 655.) Oxford: Tempus Reparatum.

Cochrane, R. A. 1931. "An Air Reconnaissance of the Hadhramaut: A Paper Read at the Evening Meeting of the Society on 15 December 1930." *The Geographical Journal* 77:209–216.

Conrad, Lawrence I. 1987. "Abraha and Muḥammad: Some Observations Apropos of Chronology and Literary *Topoi* in the Early Arabic Historical Tradition." *Bulletin of the School of Oriental and African Studies* 50:225–240.

Conrad, Lawrence I. 1988. "Seven and the *Tasbīʿ*: On the Implications of Numerical Symbolism for the Study of Medieval Islamic History." *Journal of the Economic and Social History of the Orient* 31:42–73.

Córdoba, Joaquín M. 2013. "New Perspectives about Iron Age [*sic*] and the Oasis Culture in the Oman Peninsula: Two Conclusive Seasons at al Madam (Sharjah, UAE)." *Isimu* 16:139–151.

Costa, Paolo M. 1983. "Notes on Traditional Hydraulics and Agriculture in Oman." *World Archaeology* 14:273–295.

Costa, Paolo M. 1992. "Problems of Style and Iconography in the South-Arabian Sculpture." *Yemen: Studi archeologici, storici e filologici sull'Arabia meridionale* 1:19–39.

Costa, Paolo, and Ennio Vicario. 1977. *Yemen: Land of Builders*. New York: Rizzoli.

Cowper, Alfred D. 1927. *Lime and Lime Mortars*. London: HM Stationery Office.

Dien, Mawil Izzi. 2004. *Islamic Law: From Historical Foundations to Contemporary Practice*. Notre Dame, IN: University of Notre Dame Press.

Donaldson, William J. 1996. "Observations on Measures of Capacity in Present-Day Northern Yemen." *New Arabian Studies* 3:33–49.

Donaldson, William J. 2000. *Sharecropping in the Yemen: A Study in Islamic Theory, Custom and Pragmatism*. (Studies in Islamic Law and Society, 13). Leiden: Brill.

Döring, Mathias. 2012. *In der wundersamsten Gegend der Welt: Die Phlegraeischen Felder am Golf von Neapel*. Adenstedt: Parmenios.

Dorpmüller, Sabine. 2005. *Religiöse Magie im "Buch der probaten Mittel": Analyse, kritische Edition und Übersetzung des Kitāb al-Muǧarrabāt von Muḥammad ibn Yūsuf as-Sanūsī (gest. um 895/1490)*. (Arabische Studien, 1). Wiesbaden: Harrassowitz.

Doutté, Edmond. 1909. *Magie et religion dans l'Afrique du Nord*. Alger: Jourdan.

Edens, Christopher. 2005. "Exploring Early Agriculture in the Highlands of Yemen." In *Sabaean Studies: Archaeological, Epigraphical and Historical Studies in Honour of Yūsuf M. ʿAbdallāh, Alessandro de Maigret and Christian J. Robin on the Occasion of Their Sixtieth Birthdays*, eds. Amida M. Sholan, Sabina Antonini, and Mounir Arbach, 185–211. Naples and Ṣanʿāʾ: University of Ṣanʿāʾ, Yemeni-Italian Centre for Archaeological Researches, and Centre français d'archéologie et de sciences sociales de Ṣanʿāʾ.

Edens, Christopher, and Tony J. Wilkinson. 1998. "Southwest Arabia during the Holocene: Recent Archaeological Developments." *Journal of World Prehistory* 12:55–119.

Eger, Helmut. 1984. "Rainwater Harvesting in the Yemeni Highlands: The Effect of Rainwater Harvesting on Soil Moisture Status and its Implications for Arable Farming, a Case Study of the ʿAmrān Region." In *Entwicklungsprozesse in der Arabischen Republik Jemen*, eds. Horst Kopp and Günther Schweizer, 147–169. (Jemen-Studien, 1). Wiesbaden: Reichert.

Eger, Helmut. 1987. *Runoff Agriculture: A Case Study about the Yemeni Highlands*. (Jemen-Studien, 7). Wiesbaden: Reichert.

Ekstrom, Heidi, and Christopher M. Edens. 2003. "Prehistoric Agriculture in Highland Yemen: New Results from Dhamar." *Yemen Update* 45:23–35.

El-Gawhary, Mohamed M. 1968. "Die Gottesnamen im magischen Gebrauch in den al-Būnī zugeschriebenen Werken." PhD diss., University of Bonn (unpublished).

Ellen, Roy. 2003. "Variation and Uniformity in the Construction of Biological Knowledge across Cultures." In *Nature across Cultures: Views of Nature and the Environment in Non-Western Cultures*, ed. Helaine Selin, 47–74. Dordrecht: Springer.

English, Paul W. 1968. "The Origin and Spread of Qanats in the Old World." *Proceedings of the American Philosophical Society* 112:170–181.

Eraqi-Klorman, Bat-Zion. 2009. "Yemen: Religion, Magic, and Jews." *Proceedings of the Seminar for Arabian Studies* 39:125–134.

al-Eryani, Mohamed L. 1979. "Hydrology and Ground Water Potential of the Tihama – Yemen Arab Republic." MS thesis, University of Arizona, Tuscon (unpublished).

Escher, Hermann A. 1976. *Wirtschafts- und sozialgeographische Untersuchungen in der Wâdî Mawr Region (Arabische Republik Jemen)*. (Beihefte zum Tübinger Atlas des Vorderen Orients, Reihe B [Geisteswissenschaften], 23). Wiesbaden: Reichert.

Ewald, Heinrich. 1839a. "Eine himjaritische Inschrift." *Zeitschrift für die Kunde des Morgenlandes* 2:107–109.

Ewald, Heinrich. 1839b. "Über die Sammlung arabischer und syrischer Handschriften in *British Museum*." *Zeitschrift für die Kunde des Morgenlandes* 2:190–214.

Fahd, Toufic. 1966. *La divination arabe: Études religieuses, sociologiques et folkloriques sur le milieu natif de l'Islam*. Leiden: Brill.

Fahd, Toufic. 1971. "Ḥurūf." In *The Encyclopaedia of Islam*, 2nd ed., III:595b–596b. Leiden: Brill.

Fahd, Toufic. 1978. "Istisḳā'." In *The Encyclopaedia of Islam*, 2nd ed., IV:269b–270b. Leiden: Brill.

Fahd, Toufic. 1997. "Siḥr." In *The Encyclopaedia of Islam*, 2nd ed., IX:567b–571b. Leiden: Brill.

Fedele, Francesco G. 2013. "Neolithic Settlement of the Eastern Yemen Plateau: An Exploration of Locational Choice and Land Use." *Arabian Archaeology and Epigraphy* 24:44–50.

Finster, Barbara. 1992. "An Outline of the History of Islamic Religious Architecture in Yemen." *Muqarnas* 9:124–147.

Finster, Barbara. 1993–1996. "Geschichtlicher Abriss der islamischen Sakralarchitektur im Yemen." *Eothen* 4–7: 15–32 and 182–198.

Forbes, Robert J. 1955. *Studies in Ancient Technology*. Vol. 1. Leiden: Brill.

Francaviglia, Vincenzo M. 2002. "Some Remarks on the Irrigation Systems of Ancient Yemen." In *Essays on the Late Prehistory of the Arabian Peninsula*, eds. Serge Cleuziou, Maurizio Tosi, and Juris Zarins, 111–144. (Serie orientale Roma, 93). Rome: Istituto italiano per l'Africa e l'Oriente.

Franke, Wolfgang. 1976. *Nutzpflanzenkunde*. Stuttgart: Thieme.

Frantsouzoff, Serge. 2015. "Horse-Breeding in Ancient Yemen: Some New Suggestions." In *South Arabia and its Neighbours: Phenomena of Intercultural Contacts*, ed. Iris Gerlach, 87–94. (Archäologische Berichte aus dem Jemen, 14). Wiesbaden: Reichert.

Frazer, James G. 1911. *The Golden Bough: A Study in Magic and Religion*. Part 1, *The Magic Art and the Evolution of Kings*. Vol. 1. 3rd ed. London: MacMillan.

Gardiner, Noah. 2012. "Forbidden Knowledge? Notes on the Production, Transmission, and Reception of the Major Works of Aḥmad al-Būnī." *Journal of Arabic and Islamic Studies* 12:81–142.

Geddes, Charles L. 1965. "al-D̲j̲anadī." In *The Encyclopaedia of Islam*, 2nd ed., II:441a. Leiden: Brill.

de Genouillac, Henri. 1926. *Céramique cappadocienne*. Vol. 1, *Introduction. Collection Chantre*. Paris: Geuthner.

Gerholm, Tomas. 1984. "The Making of a Traditional Society: External and Internal Factors." In *Entwicklungsprozesse in der Arabischen Republik Jemen*, eds. Horst Kopp and Günther Schweizer, 87–98. (Jemen-Studien, 1). Wiesbaden: Reichert.

Gerig, Mathias. 1982. “Beiträge zur Erforschung der antiken und mittelalterlichen Oase von Mārib.” *Archäologische Berichte aus dem Yemen* 1:33–55.

Gerlach, Iris. 2003–2004. “Die archäologisch-bauhistorischen Untersuchungen des Deutschen Archäologischen Instituts in der sabäischen Stadtanlage und Oase von Sirwāh (Jemen/Provinz Marib).” *Nürnberger Blätter zur Archäologie* 20:37–56.

Gerlach, Iris 2012. “Vorislamische Bewässerungssysteme in der Oase von Mārib, Jemen.” In *Wasserwirtschaftliche Innovationen im archäologischen Kontext: Von den prähistorischen Anfängen bis zu den Metropolen der Antike*, eds. Florian Klimscha, Ricardo Eichmann, Christof Schuler, and Henning Fahlbusch, 187–198. (Menschen – Kulturen – Traditionen: Studien aus den Forschungsclustern des Deutschen Archäologischen Instituts, 5; Forschungscluster 2: Innovationen: technisch, sozial). Rahden/Westfalen: Leidorf.

Gibb, Hamilton A. R., and Johannes H. Kramers, eds. (1953) 1974. *Shorter Encyclopaedia of Islam*. Reprint, Leiden: Brill.

Gingrich, Andre. 1993. “Regen Gottes, Land der Ehre: Konzeptualisierungen von ‘Natur’ bei den Munebbih des Nordwestjemen.” In *Metamorphosen der Natur: Sozialanthropologische Untersuchungen zum Verhältnis von Weltbild und natürlicher Umwelt*, eds. Andre Gingrich and Elke Mader, 143–159. Wien: Böhlau.

Giunta, Roberta 2002. “The Talismanic-Religious Nature of Late Ottoman Inscriptions in the Tihāmah Cities.” *Proceedings of the Seminar for Arabian Studies* 32:269–279.

Glassé, Cyril. 1989. *The Concise Encyclopaedia of Islam*. London: Stacey International.

Glennie, Kenneth W. 1970. *Desert Sedimentary Environments*. (Developments in Sedimentology, 14). Amsterdam: Elsevier.

Goblot, Henri. 1979. *Les qanats: Une technique d'acquisition de l'eau*. (Industrie et artisanat, 9). Paris: Mouton.

Goldziher, Ignaz. 1872. “Linguistisches aus der Literatur der muhammedanischen Mystik.” *Zeitschrift der Deutschen Morgenländischen Gesellschaft* 26:764–785.

Goldziher, Ignaz. 1890a. “Ueber die Entwicklung des Ḥadîth.” In *Muhammedanische Studien*, 2:1–274. Halle: Niemeyer.

Goldziher, Ignaz. 1890b. “Die Heiligenverehrung im Islam.” In *Muhammedanische Studien*, 2:275–378. Halle: Niemeyer.

Goldziher, Ignaz. 1906. “Zauberelemente im islamischen Gebet.” In *Orientalische Studien: Theodor Nöldeke zum siebzigsten Geburtstag (2. März 1906), gewidmet von Freunden und Schülern*, ed. Carl Bezold, 1:303–329. Giessen: Töpelmann.

Golvin, Lucien, and Marie-Christine Fromont. 1984. *Thulâ: Architecture et urbanisme d'une cité de haute montagne en République arabe du Yémen*. Paris: Éditions Recherche sur les civilisations.

Grohmann, Adolf. 1914. *Göttersymbole und Symboltiere auf südarabischen Denkmälern*. (Denkschriften der Kaiserlichen Akademie der Wissenschaften, Philosophisch-Historische Klasse, 58.1). Wien: Hölder.

Grohmann, Adolf. 1933. *Südarabien als Wirtschaftsgebiet*. Vol. 2. (Schriften der Philosophischen Fakultät der Deutschen Universität in Prag, 13). Brünn: Rohrer.

Guillaume, Alfred. 1955. *The Life of Muhammad: A Translation of Ibn Isḥāq's* Sīrat rasūl Allāh. Oxford: University Press.

Haddash, Salahuddin. 2001. "Water Legislation in the Republic of Yemen." In *Water in the Arabian Peninsula: Problems and Policies*, ed. Kamil A. Mahdi, 329–345. (Exeter Arab and Islamic Studies Series). Reading, UK: Ithaca.

al-Hamdānī, al-Ḥasan ibn Aḥmad/Müller, David Heinrich, ed. 1884–1891. *Al-Hamdânî's Geographie der arabischen Halbinsel nach den Handschriften von Berlin, Constantinopel, London, Paris und Strassburg*. 2 vols. Leiden: Brill.

Harrower, Michael J. 2016. *Water Histories and Spatial Archaeology: Ancient Yemen and the American West*. Cambridge: Cambridge University Press.

Hartmann-Schmitz, Ulrike. 1989. *Die Zahl Sieben im sunnitischen Islam: Studien anhand von Koran und* Ḥadīṯ. Frankfurt/Main: Lang.

Hasluck, Frederick W. 1929. *Christianity and Islam under the Sultans*. 2 vols. Oxford: Clarendon.

Hehmeyer, Ingrid. 1995. "Physical Evidence of Engineered Water Systems in Mediaeval Zabīd." *Proceedings of the Seminar for Arabian Studies* 25:45–54.

Hehmeyer, Ingrid. 2005. "Diurnal Time Measurement for Water Allocation in Southern Yemen." *Proceedings of the Seminar for Arabian Studies* 35:87–96.

Hehmeyer, Ingrid. 2007. "Water and Waste in Medieval Zabīd, Yemen" *Proceedings of the Seminar for Arabian Studies* 37:111–123.

Hehmeyer, Ingrid. 2008. "Water and Sign Magic in al-Jabīn, Yemen." *The American Journal of Islamic Social Sciences* 25 (3): 82–96.

Hehmeyer, Ingrid. 2017. "*Majil* and *Birka*: Cisterns in the Western Highlands of Yemen." In *Architectural Heritage of Yemen: Buildings That Fill My Eye*, ed. Trevor H. J. Marchand, 157–163. London: Gingko.

Hehmeyer, Ingrid, and Edward J. Keall. 1993. "Water and Land Management in the Zabid Hinterland." *Al-ʿUṣūr al-Wusṭā* 5:25–27.

Hehmeyer, Ingrid, and Jürgen Schmidt. 1991. *Antike Technologie – Die Sabäische Wasserwirtschaft von Mārib*. (Archäologische Berichte aus dem Yemen, 5). Mainz: Zabern.

Hehmeyer, Ingrid, Edward J. Keall, and Dan Rahimi. 2002. "Ghayl Bā Wazīr: Applied *Qanāt* Technology in the Fissured Karst Landscape of Southern Yemen." *Proceedings of the Seminar for Arabian Studies* 32:83–97.

Hinz, Walther. 1970. *Islamische Maße und Gewichte umgerechnet ins metrische System*. (Handbuch der Orientalistik, 1. Abteilung: Der Nahe und der Mittlere Osten, Ergänzungsband 1, Heft 1). Leiden: Brill.

Hirsch, Leo. 1897. *Reisen in Süd-Arabien, Mahra-Land und Hadramūt*. Leiden: Brill.

Hitgen, Holger. 2005a. "The Ancient Cultural Landscape of the Wādī Ǧufayna in the Oasis of Ma'rib." In *Sabaean Studies: Archaeological, Epigraphical and Historical Studies in Honour of Yūsuf M. 'Abdallāh, Alessandro de Maigret and Christian J. Robin on the Occasion of Their Sixtieth Birthdays*, eds. Amida M. Sholan, Sabina Antonini, and Mounir Arbach, 321–340. Naples and Ṣan'ā': University of Ṣan'ā', Yemeni-Italian Centre for Archaeological Researches, and Centre français d'archéologie et de sciences sociales de Ṣan'ā'.

Hitgen, Holger. 2005b. *Marib: Archäologische Touristenbroschüre*. Sanaa: Ministerium für Kultur und Tourismus der Arabischen Republik Jemen.

Ho, Engseng. 2006. *The Graves of Tarim: Genealogy and Mobility across the Indian Ocean*. Berkeley: University of California Press.

Hodge, A. Trevor. 1992. *Roman Aqueducts and Water Supply*. London: Duckworth.

Hovden, Eirik. 2006. "Rainwater Harvesting Cisterns and Local Water Management: A Qualitative Geographical/Socio-anthropological Case Study and Ethnographic Description from the Districts Hajja, Mabyan and Shiris, Governorate of Hajja, Yemen." MS thesis, University of Bergen, Norway. https://bora.uib.no/handle/1956/2001.

Hovden, Eirik. 2014. "Birka and Baraka: Cistern and Blessing." In *Southwest Arabia across History: Essays to the Memory of Walter Dostal*, eds. Andre Gingrich and Siegfried Haas, 55–65. Wien: Österreichische Akademie der Wissenschaften.

Hussain, Ghulam. 1891. "Punjab – The Jinn – Their Appearance." *North Indian Notes and Queries* 1:103, no. 678.

Ibn Baṭṭūṭa, Shams al-Dīn Abū 'Abdallāh Muḥammad ibn 'Abdallāh/Gibb, Hamilton A. R., trans. 1962. *The Travels of Ibn Baṭṭūṭa A.D. 1325–1354*. Vol. 2. Cambridge: Cambridge University Press.

Ibn al-Dayba', 'Abd al-Raḥmān ibn 'Alī/Chelhod, Joseph, ed. 1983. *Al-Faḍl al-mazīd 'alā Bughyat al-mustafīd fī akhbār madīnat Zabīd*. Ṣan'ā': Markaz al-dirāsāt wa-l-buḥūth al-yamanī.

Ibn Khallikān, Aḥmad ibn Muḥammad/de Slane, William Mac Guckin, trans. 1842. *Biographical Dictionary*. Vol. 1. Paris: Oriental Translation Fund of Great Britain and Ireland.

Ibn al-Mujāwir/Smith, G. Rex, trans. 2008. *A Traveller in Thirteenth-Century Arabia: Ibn al-Mujāwir's* Tārīkh al-mustabṣir. (The Hakluyt Society, ser. 3, 19). Aldershot, UK: Ashgate.

Ibn al-Mujāwir/Löfgren, Oscar, ed. 1951–1954. *Ibn al-Muǧāwir: Descriptio Arabiae meridionalis, praemissis capitibus de Mecca et parte regionis Ḥiǧāz; qui liber inscribitur Ta'rīḫ al-mustabṣir*. 2 vols. Leiden: Brill.

Ibn Sa'd, Muḥammad/Mittwoch, Eugen, ed. 1905. *Kitāb al-Ṭabaqāt al-kabīr*. Vol. 1, part 1, *Biographie Muhammeds bis zur Flucht*. Leiden: Brill.

Ibn Shaykhān, Sāmī M. 1999. *Nafaḥāt wa-'abīr min tārīkh Ghayl Bā Wazīr*. Aden: Maktabat al-thaqāfa; Ṣan'ā': Dār al-taysīr.

Irvine, Arthur K. 1962. "A Survey of Old South Arabian Lexical Materials Connected with Irrigation Techniques." PhD diss., University of Oxford (unpublished).

Jamme, Albert. 1962. *Sabaean Inscriptions from Maḥram Bilqîs (Mârib)*. (Publications of the American Foundation for the Study of Man, 3). Baltimore: Johns Hopkins Press.

al-Janadī, Muḥammad ibn Yūsuf ibn Yaʿqūb/al-Akwaʿ, Muḥammad ibn ʿAlī ibn al-Ḥusayn, ed. 1983. *Kitāb al-Sulūk fī ṭabaqāt al-ʿulamāʾ wa-l-mulūk*. Al-Jumhūriyya al-ʿarabiyya al-yamaniyya: Wizārat al-aʿlām wa-l-thaqāfa.

Jashemski, Wilhelmina F. 1979. *The Gardens of Pompeii*. New Rochelle, NY: Caratzas.

Jāzim, Muḥammad ʿAbd al-Raḥīm, ed. 2003–2005. *Nūr al-maʿārif fī nuẓum wa-qawānīn wa-aʿrāf al-Yaman fī l-ʿahd al-muẓaffarī al-wārif/Lumière de la connaissance: Règles, lois et coutumes du Yémen sous le règne du sultan rasoulide al-Muẓaffar*. 2 vols. Ṣanʿāʾ: al-Maʿhad al-faransī lil-āthār wa-l-ʿulūm al-ijtimāʿiyya bi-Ṣanʿāʾ/Centre français d'archéologie et de sciences sociales.

Jāzim, Muḥammad ʿAbd al-Raḥīm, ed. 2008. *Irtifāʿ al-dawla al-Muʾayyadiyya: Jibāyat bilād al-Yaman fī ʿahd al-sulṭān al-Malik al-Muʾayyad Dāwūd b. Yūsuf al-Rasūlī (al-mutawaffā sanat 721 h/1321 m)/Le Livre des Revenus du sultan rasûlide al-Malik al-Muʾayyad Dâwûd b. Yûsuf (m. 721/1321)*. (La bibliothèque yéménite/al-Maktaba al-yamaniyya/Die Jemenitische Bibliothek, 2). Ṣanʿāʾ: al-Maʿhad al-faransī lil-āthār wa-l-ʿulūm al-ijtimāʿiyya bi-Ṣanʿāʾ, al-Maʿhad al-almānī lil-āthār, Ṣanʿāʾ/Centre français d'archéologie et de sciences sociales, Deutsches Archäologisches Institut, Sanaa Branch.

Kaḥḥāla, ʿUmar R. 1957–1961. *Muʿjam al-muʾallifīn*. 15 vols. Damascus: al-Maktaba al-ʿarabiyya.

Kamali, Mohammad H. 2008. *Shariʿah Law: An Introduction*. Oxford: Oneworld.

Keall, Edward J. 1983. "Zabīd and its Hinterland: 1982 Report." *Proceedings of the Seminar for Arabian Studies* 13:53–69.

Keall, Edward J. 1984. "A Preliminary Report on the Architecture of Zabīd." *Proceedings of the Seminar for Arabian Studies* 14:51–65.

Keall, Edward J. 1989a. "A Few Facts about Zabīd." *Proceedings of the Seminar for Arabian Studies* 19:61–69.

Keall, Edward J. 1989b. "Revealing the Past through Reconstructed Walls." *Rotunda* 22 (2): 28–33.

Keall, Edward J. 1991. "Drastic Changes in 16th Century Zabīd." *Proceedings of the Seminar for Arabian Studies* 21:79–96.

Keall, Edward J. 1992. "Smokers' Pipes and the Fine Pottery Tradition of Hays." *Proceedings of the Seminar for Arabian Studies* 22:29–46.

Keall, Edward J. 1993. "The Medieval Environment of Zabid, Yemen." *National Geographic Research and Exploration* 9 (4): 475–477.

Keall, Edward J. 2001a. "The Evolution of the First Coffee Cups in Yemen." In *Le commerce du café avant l'ère des plantations coloniales: Espaces, réseaux, sociétés*

(*XVe-XIXe siècle*), ed. Michel Tuchscherer, 35–50. (Cahier des annales islamologiques, 20). Cairo: Institut français d'archéologie orientale.

Keall, Edward J. 2001b. "The Syrian Origins of Yemen's National Mosque Style." *Bulletin of the Canadian Society for Mesopotamian Studies* 37:219–226.

Keall, Edward J. 2012. "Getting to the Bottom of Zabid: The Canadian Archaeological Mission in Yemen, 1982–2011." *Proceedings of the Seminar for Arabian Studies* 42:129–142.

Keall, Edward J., and Ingrid Hehmeyer. 1998. "Sponsorship of a *Madrasa*, Reflecting the Value of Farmland in the Urban Economy of Zabīd, Yemen." *Al-ʿUṣūr al-Wusṭā* 10:33–35 and 47.

Kenderova, Stoyanka. 2014. "Une chronique de Zabid et du Yémen jusqu'en 1562: Le manuscrit OR. 2545 de Sofia (Bulgarie). *Journal of Islamic Manuscripts* 5:170–197.

al-Khazrajī, ʿAlī ibn al-Ḥasan/Redhouse, James W., trans., and ʿAsal, Muḥammad, ed. 1906–1918. *The Pearl-Strings; A History of the Resúliyy Dynasty of Yemen, by ʿAliyyu ʾbnu ʾl-Ḥasan ʾel-Khazrejiyy*. 5 vols. (E. J. W. Gibb Memorial Series, 3). Leiden: Brill; London: Luzac.

King, David A. 1990. "Science in the Service of Religion: The Case of Islam." *Impact of Science on Society* 159:245–262.

King, David A. 1996. "Astronomy and Islamic Society: Qibla, Gnomonics and Timekeeping." In *Encyclopedia of the History of Arabic Science*, ed. Roshdi Rashed, 1:128–184. London: Routledge.

King, David A. 2012. "The Invention of Algebra in Zabīd: Between Legend and Fact." In *Islamic Philosophy, Science, Culture, and Religion: Studies in Honor of Dimitri Gutas*, eds. Felicitas Opwis and David Reisman, 223–231. Leiden: Brill.

Klessing, Jan M. 1995. "Planung und Ausführung der Restaurierung." In *Sana'a: Die Restaurierung der Samsarat al-Mansurah/The Restoration of the Samsarat al-Mansurah*, eds. Michael Petzet and Wolf Koenigs, 49–91. (Arbeitshefte des Bayerischen Landesamtes für Denkmalpflege, 70). Munich: Lipp.

König, Eduard. 1907. "Die Zahl vierzig und Verwandtes." *Zeitschrift der Deutschen Morgenländischen Gesellschaft* 61:913–917.

Kopp, Horst. 1981. *Agrargeographie der Arabischen Republik Jemen*. (Erlanger Geographische Arbeiten, Sonderband 11). Erlangen: Fränkische Geographische Gesellschaft.

Kopp, Horst, ed. 2005. *Länderkunde Jemen*. Wiesbaden: Reichert.

Kopp, Horst. 2008. "Landesnatur und kulturelle Vielfalt des Jemen – Eine Einführung." In *Jemen – kulturgeographische Erkundungen*, ed. Konrad Schliephake, 137–142. (Würzburger Geographische Manuskripte, 74). Würzburg: Institut für Geographie der Universität Würzburg.

Kriss, Rudolf, and Hubert Kriss-Heinrich. 1960–1962. *Volksglaube im Bereich des Islam*. 2 vols. Wiesbaden: Harrassowitz.

Kropp, Manfred. 1994. "'... nur zum Abschluß planmäßiger Bauarbeiten, keinesfalls als Katastrophenhilfe!' Das politische Management eines Dammbruchs bei Marib durch König Šaraḫbi'il Ya'fur (CIH 540,64–72)." In *Arabia Felix: Beiträge zur Sprache und Kultur des vorislamischen Arabien. Festschrift Walter W. Müller zum 60. Geburtstag*, ed. Norbert Nebes, 123–131. Wiesbaden: Harrassowitz.

al-Kūfī, Muḥammad ibn Sulaymān. 1993. *Kitāb al-Muntakhab*. Ṣanʿāʾ: Dār al-ḥikma al-yamāniyya.

Kusaybī, Nazīh. 1985. *Fihris al-makhṭūṭāt al-ʿarabiyya fī maktabat Strāsbūrgh al-waṭaniyya wa-l-jāmiʿiyya/Catalogue critique des manuscrits arabes de la Bibliothèque nationale et universitaire de Strasbourg*. Manshūrāt maʿhad al-makhṭūṭāt al-ʿarabiyya/Publications de l'Institut des manuscrits arabes. Kuwait: al-Munaẓẓama al-ʿarabiyya lil-tarbiya wa-l-thaqāfa wa-l-ʿulūm/Organisation de la ligue arabe pour l'éducation, la culture et les sciences.

Laessøe, Jørgen. 1951. "The Irrigation System at Ulḫu, 8th Century B.C." *Journal of Cuneiform Studies* 5:21–32.

Lambton, Ann K. S. 1978. "Ḳanāt. I. In Iran." In *The Encyclopaedia of Islam*, 2nd ed., IV:528b–532a. Leiden: Brill.

Lambton, Ann K. S. 1989. "The Origin, Diffusion and Functioning of the Qanat." In *Qanat, Kariz and Khattara: Traditional Water Systems in the Middle East and North Africa*, eds. Peter Beaumont, Michael Bonine, and Keith McLachlan, 5–12. London: School of Oriental and African Studies.

Lane, Edward William. 1863–1893. *An Arabic-English Lexicon*. 8 vols. London: Williams and Norgate.

Laureano, Pietro. 2003. "A History of Traditional Water Management." In *Sustainable Management of Marginal Drylands: Applications of Indigenous Knowledge for Coastal Drylands*, ed. Zafar Adeel, 127–146. Tokyo: United Nations University.

Lee Warner, W. H. 1931. "Notes on the Hadhramaut: Expanded from His Contribution to the Discussion on the Preceding Paper." *The Geographical Journal* 77:217–222.

Lewcock, Ronald. 1983. "The Buildings of the Sūq/Market." In *Ṣanʿāʾ: An Arabian Islamic City*, eds. Robert B. Serjeant and Ronald Lewcock, 276–302. London: World of Islam Festival Trust.

Lewcock, Ronald, and Robert B. Serjeant. 1983. "The Houses of Ṣanʿāʾ." In *Ṣanʿāʾ: An Arabian Islamic City*, eds. Robert B. Serjeant and Ronald Lewcock, 436–500. London: World of Islam Festival Trust.

Libson, Gideon. 2003. *Jewish and Islamic Law: A Comparative Study of Custom during the Geonic Period*. Cambridge, MA: Harvard University Press.

Lichtenthäler, Gerhard. 2000. "Power, Politics and Patronage: Adaptation of Water Rights among Yemen's Northern Highland Tribes." *Études rurales* 155–156:143–166.

Lightfoot, Dale. 2000a. "Ghayl and Miyan in Arabia Felix: The Ecology of Diffusion and Recession of Use." *The Arab World Geographer* 3:2–21.

Lightfoot, Dale. 2000b. "The Origin and Diffusion of *Qanats* in Arabia: New Evidence from the Northern and Southern Peninsula." *The Geographical Journal* 166:215–226.

Little, Otway H. 1925. *The Geography and Geology of Makalla (South Arabia)*. (Survey of Egypt: Geological Survey). Cairo: Government Press.

Livingston, Morna. 2002. *Steps to Water: The Ancient Stepwells of India*. New York: Princeton Architectural Press.

Lucas, A. 1944. "Report on Mortars and Plasters from the Temple: Memorandum." In Gertrude Caton Thompson, *The Tombs and Moon Temple of Hureida (Hadhramaut)*, 62. (Reports of the Research Committee of the Society of Antiquaries of London, 13). Oxford: Oxford University Press; London: The Society of Antiquaries.

Lüttge, Ulrich, Manfred Kluge, and Gerhard Thiel. 2010. *Botanik: Die umfassende Biologie der Pflanzen*. Weinheim: Wiley-VCH.

Macdonald, Duncan B. 2004. "Budūḥ." In *The Encyclopaedia of Islam*, 2nd ed., XII:153a–154a. Leiden: Brill.

Macdonald, Michael C. A. 2000. "Reflections on the Linguistic Map of Pre-Islamic Arabia." *Arabian Archaeology and Epigraphy* 11:28–79.

Maktari, Abdullah M. A. 1971. *Water Rights and Irrigation Practices in Laḥj: A Study of the Application of Customary and* Sharīʿah *Law in South-West Arabia*. (Oriental Publications, 21). Cambridge: Cambridge University Press.

Maraqten, Mohammed. 1993. "Wine Drinking and Wine Prohibition in Arabia before Islam." *Proceedings of the Seminar for Arabian Studies* 23:95–115.

Maraqten, Mohammed. 2014. *Altsüdarabische Texte auf Holzstäbchen: Epigraphische und kulturhistorische Untersuchungen*. (Beiruter Texte und Studien, 103). Würzburg: Ergon.

Marchand, Trevor H. J. 2012. *Minaret Building and Apprenticeship in Yemen*. London: Routledge.

Maréchaux, Pascal. 1980. *Arabia Felix: Images of Yemen and its People*. Woodbury, NY: Barron's Educational Series.

al-Māwardī, ʿAlī ibn Muḥammad/Yate, Asadullah, trans. 1996. *Al-Ahkam as-Sultaniyyah: The Laws of Islamic Governance*. London: Ta-Ha.

Mazzini, Giovanni, and Alexandra Porter. 2009. "Stela BM 102600=CIH 611 in the British Museum: Water Regulation between Two Bordering Estates." *Proceedings of the Seminar for Arabian Studies* 39:283–294.

van der Meulen, Daniel. 1932. "Into Burning Hadhramaut. The Arab Land of Frankincense and Myrrh, Ever a Lodestone of Western Exploration." *The National Geographic Magazine* 62:387–429.

van der Meulen, Daniel, and Hermann von Wissmann. 1932. *Ḥaḍramaut: Some of its Mysteries Unveiled*. Leiden: Brill.

Meyerhof, Max. 1944. "Sur un traité d'agriculture composé par un sultan yéménite du XIVe siècle (deuxième partie)." *Bulletin de l'Institut d'Égypte* 26:51–65.

Miles, S. B., and M. W. Munzinger. 1871. "Account of an Excursion into the Interior of Southern Arabia." *The Journal of the Royal Geographical Society* 41:210–245.

Milstein, Rachel. n.d. [1995]. *King Solomon's Seal*. Jerusalem: Tower of David Museum of the History of Jerusalem.

Milwright, Marcus. 2010. *An Introduction to Islamic Archaeology*. Edinburgh: Edinburgh University Press.

Mols, Luitgard E. M. 2006. *Mamluk Metalwork Fittings in Their Artistic and Architectural Context*. Delft: Eburon.

Mordtmann, Andreas. 1864. "Studien über geschnittene Steine mit Pehlevi-Inschriften." *Zeitschrift der Deutschen Morgenländischen Gesellschaft* 18:1–52.

de Moulins, Dominique, Carl Phillips, and Nadia Durrani. 2003. "The Archaeobotanical Record of Yemen and the Question of Afro-Asian Contacts." In *Food, Fuel and Fields: Progress in African Archaeobotany*, eds. Katharina Neumann, Ann Butler, and Stefanie Kahlheber, 213–228. Cologne: Heinrich-Barth-Institut.

Müller, David H. 1899. *Südarabische Alterthümer im kunsthistorischen Hofmuseum*. Wien: Hölder.

Müller, Walter W. 1983. "Altsüdarabische Dokumente." In *Texte aus der Umwelt des Alten Testaments*, ed. Otto Kaiser, 1 (3): 268–282. Gütersloh: Mohn.

Müller, Walter W. 1985. "Altsüdarabische und frühnordarabische Inschriften." In *Texte aus der Umwelt des Alten Testaments*, ed. Otto Kaiser, 1 (6): 651–668. Gütersloh: Mohn.

Müller, Walter W. 1986. "Eine Bitte um Regen bei Dürre aus sabäischer Zeit." *Jemen-Report* 17 (1): 10–11.

Müller, Walter W. 1987. "Skizze der Geschichte Altsüdarabiens." In *Jemen: 3000 Jahre Kunst und Kultur des glücklichen Arabien*, ed. Werner Daum, 50–56. Innsbruck: Pinguin; Frankfurt/Main: Umschau.

Müller, Walter W. 1988a. "Altsüdarabische Rituale und Beschwörungen." *Texte aus der Umwelt des Alten Testaments*, ed. Otto Kaiser, 2 (3): 438–452. Gütersloh: Mohn.

Müller, Walter W. 1988b. "Altsüdarabische und frühnordarabische Grab-, Sarkophag-, Votiv- und Bauinschriften." In *Texte aus der Umwelt des Alten Testaments*, ed. Otto Kaiser, 2 (4): 621–640. Gütersloh: Mohn.

Müller, Walter W. 1991. "Mārib." In *The Encyclopaedia of Islam*, 2nd ed., VI:559a–567b. Leiden: Brill.

Müller, Walter W. 1999. "Die Stele des ʾAbraha, des äthiopischen Königs im Jemen." In *Im Land der Königin von Saba: Kunstschätze aus dem antiken Jemen. 7. Juli 1999–9. Januar 2000*, ed. Staatliches Museum für Völkerkunde München, 268–270. Germering/München: International Publishing.

Munro, R. Neil, and Tony J. Wilkinson. 2007. "Environment, Landscapes and Archaeology in the Yemeni Tihāmah." In *Natural Resources and Cultural Connections of the Red*

Sea, eds. Janet Starkey, Paul Starkey, and Tony J. Wilkinson, 13–33. (BAR International Series, 1661). Oxford: Archaeopress.

al-Muqaddasī, Shams al-Dīn Muḥammad ibn Aḥmad/Collins, Basil Anthony, trans. 1994. *The Best Divisions for Knowledge of the Regions: A Translation of* Ahsan al-Taqasim fi Maʿrifat al-Aqalim. Reading, UK: Garnet.

Muscarella, Oscar W. 1986. "The Location of Ulhu and Uiše in Sargon II's Eighth Campaign, 714 B.C." *Journal of Field Archaeology* 13:465–475.

al-Nahrawālī, Quṭb al-Dīn Muḥammad ibn Aḥmad/al-Jāsir, Ḥamad, ed. 1967. *Al-Barq al-yamānī fī l-fatḥ al-ʿuthmānī*. Riyadh: Dār al-yamāma lil-baḥth wa-l-tarjama wa-l-nashr.

National Palace Museum. 1974. *Illustrated Catalogue of Sung Dynasty Porcelain. Lung-ch'üan Ware, Ko Ware and Other Wares*. Taipei, Taiwan: National Palace Museum.

Naval Intelligence Division. 1946. *Western Arabia and the Red Sea*. (Geographical Handbook Series, B.R. 527 [restricted; for official use only]). Oxford: Naval Intelligence Division.

Nebes, Norbert. 1996. "Weihrauch und Myrrhe der Königin von Saba: Zur Entstehung früher Hochkulturen in Südarabien." *Wirtschaft und Wissenschaft* 4:13–22.

Nebes, Norbert. 2001. "Zur Genese der altsüdarabischen Kultur: Eine Arbeitshypothese." In *Migration und Kulturtransfer: Der Wandel vorder- und zentralasiatischer Kulturen im Umbruch vom 2. zum 1. vorchristlichen Jahrtausend. Akten des Internationalen Kolloquiums Berlin, 23. bis 26. November 1999/Eurasien- und Orient-Abteilung des Deutschen Archäologischen Instituts*, eds. Ricardo Eichmann and Hermann Parzinger, 427–435. Bonn: Habelt.

Nebes, Norbert. 2004. "A New ʾAbraha Inscription from the Great Dam of Mārib." *Proceedings of the Seminar for Arabian Studies* 34:221–230.

Nebes, Norbert. 2012. "Südarabien." In *Reallexikon der Assyriologie und Vorderasiatischen Archäologie* 13 (3–4): 243b–248a. Berlin: De Gruyter.

Niebuhr, Carsten. (1772) 1969. *Beschreibung von Arabien aus eigenen Beobachtungen und im Lande selbst gesammelten Nachrichten*. Reprint, Graz: Akademische Druck- und Verlagsanstalt.

Niebuhr, Carsten. (1774) 1968. *Reisebeschreibung nach Arabien und den umliegenden Ländern*. Vol. 1. Reprint, Graz: Akademische Druck- und Verlagsanstalt.

Nöldeke, Theodor. 1898. "Zur tendenziösen Gestaltung der Urgeschichte des Islāms." *Zeitschrift der Deutschen Morgenländischen Gesellschaft* 52:16–33.

Nützel, Heinrich. 1892. "Münzen der Rasuliden nebst einem Abriss der Geschichte dieser jemenitischen Dynastie." *Zeitschrift für Numismatik* 18:81–156.

Paret, Rudi. 1958. *Symbolik des Islam*. (Symbolik der Religionen, 2). Stuttgart: Hiersemann.

Peters, Ruud. 2002. "Waḳf. I. In Classical Islamic Law." In *The Encyclopaedia of Islam*, 2nd ed., XI: 59a–63b. Leiden: Brill.

Pielow, Dorothee A. M. 1995. *Die Quellen der Weisheit: Die arabische Magie im Spiegel des Uṣūl al-Ḥikma von Aḥmad Ibn ʿAlī al-Būnī.* (Arabistische Texte und Studien, 8). Hildesheim: Olms.

Polybius/Paton, William R., trans. 1922–1927. *The Histories.* 6 vols. (The Loeb Classical Library). London: Heinemann; New York: Putnam's Sons.

Porter, Venetia A. 1992. "The History and Monuments of the Tahirid Dynasty of the Yemen 858–923/1454–1517." PhD diss., University of Durham (unpublished).

Porter, Venetia. 1997. "The Architectural Decoration of the ʿAmiriya. Painting, Stucco and Qudad." In Selma Al-Radi, *The ʿAmiriya in Radaʿ: The History and Restoration of a Sixteenth-Century Madrasa in the Yemen*, 113–138. (Oxford Studies in Islamic Art, 13). Oxford: Oxford University Press.

Porter, Venetia. 1998. "Islamic Seals: Magical or Practical?" In *University Lectures in Islamic Studies*, ed. Alan Jones, 2:135–149. London: Altajir World of Islam Trust.

Potts, Daniel T. 1990. *The Arabian Gulf in Antiquity.* Vol. 1, *From Prehistory to the Fall of the Achaemenid Empire.* Oxford: Clarendon.

Pritzkat, Thomas. 1999. "'Marktwirtschaft' statt Sozialismus in Südarabien: Die Reform der Bodenordnung im Südjemen am Beispiel Mukalla." *Jemen-Report* 30 (2): 25–31.

Rappold, Gerhard D., Peter Ergenzinger, Horst Gerke, and Fred Scholz. 2003. "Hydrological Analysis of Terraced Catchments: A Case Study of the Taʿizz Region, Yemen." In *Indigenous Knowledge and Sustainable Agriculture in Yemen*, eds. Amin al-Hakimi and Frédéric Pelat, 71–78. (Les cahiers du Centre français d'archéologie et de sciences sociales de Sanaa, 3). Sanaa: Centre français d'archéologie et de sciences sociales.

Ransom, Marjorie. 2014. *Silver Treasures from the Land of Sheba: Regional Yemeni Jewelry.* Cairo: The American University Press.

Rasmussen, Stig T. 1996. "Journeys in Persia and Arabia in the 17th and 18th Centuries." In *The Arabian Journey: Danish Connections with the Islamic World over a Thousand Years*, 45–65. Århus: Prehistoric Museum Moesgård.

Rathjens, Carl, and Hermann von Wissmann. 1932. *Vorislamische Altertümer.* (Rathjens v. Wissmannsche Südarabien-Reise, 2). Hamburg: Friederichsen, De Gruyter.

Rathjens, Carl, and Hermann von Wissmann. 1934. *Landeskundliche Ergebnisse.* (Rathjens-v. Wissmannsche Südarabien-Reise, 3). Hamburg: Friederichsen, De Gruyter.

Regourd, Anne. 2007. "Deux coupes magico-thérapeutiques, biens de fondation pieuse (Nord du Yémen): Transmission du savoir et efficacité." In *Coran et talismans: Textes et pratiques magiques en milieu musulman*, ed. Constant Hamès, 309–345. Paris: Karthala.

Reinaud, Joseph T. 1828. *Monuments arabes, persans et turcs, du cabinet de M. le Duc de Blacas et d'autres cabinets.* 2 vols. Paris: L'imprimerie royale.

Remmele, Gerhard. 1989. "Die Niederschlagsverhältnisse im Südwesten der arabischen Halbinsel." *Erdkunde* 43:27–36.

Rhodokanakis, Nikolaus. 1931. "Zur altsüdarabischen Epigraphik und Archäologie I." *Wiener Zeitschrift für die Kunde des Morgenlandes* 38:167–182.

Riesebrodt, Martin. 2008. "Theses on a Theory of Religion." *International Political Anthropology* 1:25–41.

Robin, Christian. 1996. "Sheba. II. Dans les inscriptions d'Arabie du Sud." In *Dictionnaire de la Bible*, supplément 12, 1047–1254. Paris: Letouzey & Ané.

Rodionov, Mikhail. 1997. "Mawlā Maṭar and Other *Awliyā'*: On Social Functions of Religious Places in Western Ḥaḍramawt." *Mare Erythraeum* 1:107–114.

Rodionov, Mikhail. 1999. "Irrigation in Western Ḥaḍramawt: *Khayyil* as a Social Role." *Proceedings of the Seminar for Arabian Studies* 29:119–121.

Rodionov, Mikhail, and Hanne Schönig. 2011. *The Hadramaut Documents, 1904–51: Family Life and Social Customs under the Last Sultans*. (Beiruter Texte und Studien, 130). Beirut: Orient-Institut; Würzburg: Ergon.

Ron, Zvi Y. D. 1989. "Qanats and Spring Flow Tunnels in the Holy Land." In *Qanat, Kariz and Khattara: Traditional Water Systems in the Middle East and North Africa*, eds. Peter Beaumont, Michael Bonine, and Keith McLachlan, 210–236. London: School of Oriental and African Studies.

Ryckmans, Jacques. 1966. "Himyaritica (2)." *Le Muséon* 79:475–500.

Ryckmans, Jacques. 1973. "Un rite d'istisqâ' au temple sabéen de Mârib." *Annuaire de l'Institut de Philologie et d'Histoire Orientales et Slaves* 20:379–388.

Ryckmans, Jacques. 1983. "Biblical and Old South Arabian Institutions: Some Parallels." In *Arabian and Islamic Studies: Articles Presented to R. B. Serjeant on the Occasion of His Retirement from the Sir Thomas Adams's Chair of Arabic at the University of Cambridge*, eds. Robin L. Bidwell and G. Rex Smith, 14–25. London: Longman.

Sabra, Abdelhamid I. 1971. "ʿIlm al-ḥisāb." In *The Encyclopaedia of Islam*, 2nd ed., III: 1138a–1141b. Leiden: Brill.

Sadek, Noha. 1989. "Rasūlid Women: Power and Patronage." *Proceedings of the Seminar for Arabian Studies* 19:121–136.

Sadek, Noha. 1990. "Patronage and Architecture in Rasūlid Yemen: 626–858 A.H./1229–1454 A.D." PhD diss., University of Toronto (unpublished).

Sadek, Noha. 2002a. "Zabīd." In *The Encyclopaedia of Islam*, 2nd ed., XI:370a–371b. Leiden: Brill.

Sadek, Noha. 2002b. "Zabīd: The Round City of Yemen." In *Studies on Arabia in Honour of Professor G. Rex Smith*, eds. John F. Healey and Venetia Porter, 215–226. (Journal of Semitic Studies, Supplement 14). Oxford: Oxford University Press.

Sait, Siraj, and Hilary Lim. 2006. *Land, Law and Islam: Property and Human Rights in the Muslim World*. London: Zed Books.

Salameh, Nadim H. 1995. "Šarīʿa und Gewohnheitsrecht in Zabīd: dargestellt an ausgewählten Fällen zum Wasserrecht aus der Lehrschrift des Muḥammad ibn Ziyād al-Waḍḍāḥī." MA thesis, Johannes Gutenberg-Universität Mainz, Germany (unpublished).

Salameh, Nadim H. 1999. "Customary Water-Rights in Mediaeval Wādī Zabīd: Some Legal Cases on *al-ʿādil bi'l-qanāʿah*." *Proceedings of the Seminar for Arabian Studies* 29:137–142.

Salameh, Nadim H. 2001. "Water Rights in 16th Century Ḥaḍramawt – One Question, Two Answers. The Legal Opinions of Ibn Ḥağar al-Haytamī and al-Bakrī al-Ṣiddīqī." *Proceedings of the Seminar for Arabian Studies* 31:215–220.

Savage-Smith, Emilie. 1997. "Amulets and Related Talismanic Objects." In Francis Maddison and Emilie Savage-Smith, *Science, Tools and Magic*. Vol. 1, *Body and Spirit, Mapping the Universe*, 132–147. (The Nasser D. Khalili Collection of Islamic Art, 12.1). London and Oxford: The Nour Foundation, in association with Azimuth Editions and Oxford University Press.

Savage-Smith, Emilie. 2004. Introduction to *Magic and Divination in Early Islam*, ed. Emilie Savage-Smith, xiii–li. (The Formation of the Classical Islamic World, 42). Aldershot, UK: Ashgate.

Schaloske, Michael. 1995. *Antike Technologie – Die Sabäische Wasserwirtschaft von Mārib*. Teil 3, *Untersuchungen der Sabäischen Bewässerungsanlagen in Mārib*. (Archäologische Berichte aus dem Yemen, 7). Mainz: Zabern.

Schmitz, Charles. 2013. "Geography." In *Yemen*, ed. Steven C. Caton, 1–27. (Middle East in Focus). Santa Barbara, CA: ABC-CLIO.

Schoenig, Hanne. 2006. "The Manifold Use of the Jujube Tree (*Ziziphus spina-christi*) in Ḥaḍramawt and Other Parts of Yemen. Some Ethnobotanical and Philological Remarks." In *The Culture of Arabia in the Asian Context: In Honor of Mikhail Rodionov's 60th Anniversary*, 72–85. St. Petersburg: Russian Academy of Sciences, Peter the Great Museum of Anthropology and Ethnography (Kunstkammer).

Schultz, Warren C. 2003. "Mamlūk Metrology and the Numismatic Evidence." *Al-Masāq* 15:59–75.

Sedov, Alexander V., and Ahmad Bâtâyiʿ. 1994. "Temples of Ancient Hadramawt." *Proceedings of the Seminar for Arabian Studies* 24:183–196.

Seipel, Wilfried, ed. 1998. *Jemen: Kunst und Archäologie im Land der Königin von Sabaʾ*. Wien: Kunsthistorisches Museum; Milano: Skira.

Semsar Yazdi, Ali A., and Majid Labbaf Khaneiki. 2010. *Veins of Desert: A Review of the Technique of Qanat/Falaj/Karez*. Tehran: Iran Water Resources Management Organization (IWRMO) and International Centre on Qanats and Historic Hydraulic Structures (ICQHS, under the auspices of UNESCO).

Serjeant, Robert B. 1964. "Some Irrigation Systems in Ḥaḍramawt." *Bulletin of the School of Oriental and African Studies* 27:33–76.

Serjeant, Robert B. 1974. "The Cultivation of Cereals in Medieval Yemen. (A Translation of the *Bughyat al-Fallāḥīn* of the Rasūlid Sultan, al-Malik al-Afḍal al-ʿAbbās b. ʿAlī, Composed circa 1370 A.D.)." *Arabian Studies* 1:25–74.

Serjeant, Robert B. 1976. *South Arabian Hunt*. London: Luzac.

Serjeant, Robert B. 1979. "Customary Law Documents as a Source of History." In *Studies in the History of Arabia*. Vol. 1, *Sources for the History of Arabia*. Part 2, 99–103. Riyad: University Press.

Serjeant, Robert B. 1983a. "The Post Medieval and Modern History of Ṣanʿāʾ and the Yemen, ca. 953–1382/1515–1962." In *Ṣanʿāʾ: An Arabian Islamic City*, eds. Robert B. Serjeant and Ronald Lewcock, 68–107 London: World of Islam Festival Trust.

Serjeant, Robert B. 1983b. "The Mosques of Ṣanʿāʾ: The Yemeni Islamic Setting." In *Ṣanʿāʾ: An Arabian Islamic City*, eds. Robert B. Serjeant and Ronald Lewcock, 310–322. London: World of Islam Festival Trust.

Serjeant, Robert B. 1988. "Observations on Irrigation in South West Arabia." *Proceedings of the Seminar for Arabian Studies* 18:145–153.

Serjeant, Robert B., Paolo Costa, and Ronald Lewcock. 1983. "The *Ghayls* of Ṣanʿāʾ." In *Ṣanʿāʾ: An Arabian Islamic City*, eds. Robert B. Serjeant and Ronald Lewcock, 19–31. London: World of Islam Festival Trust.

Shaqliyya, Aḥmad R. 1992. "Al-Kharīṭa al-zirāʿiyya al-muʿāṣira lil-bilād al-yamaniyya." *Dirāsāt yamaniyya* 44:395–423.

Shatzmiller, Maya. 2001. "Islamic Institutions and Property Rights: The Case of the 'Public Good' Waqf." *Journal of the Economic and Social History of the Orient* 44:44–74.

Sima, Alexander 2000. *Tiere, Pflanzen, Steine und Metalle in den altsüdarabischen Inschriften*. (Akademie der Wissenschaften und der Literatur, Mainz, Veröffentlichungen der Orientalischen Kommission, 46). Wiesbaden: Harrassowitz.

Smith, G. Rex. 1969. "The Ayyūbids and Rasūlids – the Transfer of Power in 7th/13th Century Yemen." *Islamic Culture* 43:175–188.

Smith, G. Rex. 1978. *The Ayyūbids and Early Rasūlids in the Yemen (567–694/1173–1295)*. Vol. 2, *A Study of Ibn Ḥātim's* Kitāb al-Simṭ *Including Glossary, Geographical and Tribal Indices and Maps*. (E. J. W. Gibb Memorial, n.s., 26.2). London: Luzac.

Smith, G. Rex. 1987. "The Political History of the Islamic Yemen down to the First Turkish Invasion (1–945/622–1538). In *Yemen: 3000 Years of Art and Civilization in Arabia Felix*, ed. Werner Daum, 129–139. Innsbruck: Pinguin; Frankfurt/Main: Umschau.

Smith, G. Rex. 1996. "The Language of Ibn al-Mujāwir's 7th/13th Century Guide to Arabia, *Tārīkh al-mustabṣir*." In *Tradition and Modernity in Arabic Language and Literature*, ed. Jack R. Smart, 327–351. Richmond, Surrey: Curzon.

Smith, G. Rex. 2002. "Al-Yaman. 3. History. (a) From Pre-Islamic Times to 1962." In *The Encyclopaedia of Islam*, 2nd ed., XI: 271b–274b.

Snow, John. 1849. "On the Pathology and Mode of Communication of Cholera." *London Medical Gazette* 44:745–752 and 923–929.

Stein, Peter. 2010a. *Die altsüdarabischen Minuskelinschriften auf Holzstäbchen aus der Bayerischen Staatsbibliothek in München*. Vol. 1, *Die Inschriften der mittel- und spätsabäischen Periode*. (Epigraphische Forschungen auf der Arabischen Halbinsel, 5). Tübingen: Wasmuth.

Stein, Peter. 2010b. "Irrigation Management in Pre-Islamic South Arabia according to the Epigraphic Evidence." *Proceedings of the Seminar for Arabian Studies* 40:337–344.

Stookey, Robert W. 1978. *Yemen: The Politics of the Yemen Arab Republic*. Boulder, CO: Westview.

Strasburger, Eduard. 1978. *Lehrbuch der Botanik für Hochschulen*. 31st ed. Stuttgart: Fischer.

Sutter, Anita. 1999. "Qadad: Seine Geschichte und seine Herstellung." *Jemen-Report* 30 (1): 12–15.

Swagman, Charles F. 1988. *Development and Change in Highland Yemen*. Salt Lake City: University of Utah Press.

Tesco-Viziterv-Vituki. 1971. "Survey of the Agricultural Potential of the Wadi Zabid, Yemen Arab Republic: Land Tenure and Water Rights." Report prepared for the Food and Agriculture Organization of the United Nations acting as the executing agency for the United Nations Development Programme. (AGL: SF/YEM 1, Technical Report, 9). Budapest (unpublished).

Thureau-Dangin, François. 1912. *Une relation de la huitième campagne de Sargon (714 av. J.-C.): Texte Assyrien inédit, publié et traduit*. Paris: Geuthner.

Tihama Development Authority. n.d. "Wadi Development for Agriculture in Yemen Arab Republic." In *Country Papers*, 2:26–33. Hodayda, Yemen Arab Republic: Tihama Development Authority (unpublished).

van der Tuin, Johannes D. 1987. *Elsevier's Dictionary of Water and Hydraulic Engineering in Five Languages: English, French, Spanish, Dutch, and German*. Amsterdam: Elsevier.

Ullmann, Manfred. 1978. "K͟hāṣṣa." In *The Encyclopaedia of Islam*, 2nd ed., IV:1097b–1098a. Leiden: Brill.

ʿUmāra al-Yamanī, Najm al-Dīn/Kay, Henry Cassels, ed. and trans. (1892) 1968. *Yaman, its Early Mediaeval History, by Najm ad-Din ʿOmārah al-Ḥakami; also the Abridged History of its Dynasties by Ibn Khaldūn and an Account of the Karmathians of Yaman by Abu ʿAbd Allah Baha ad-Din al-Janadi*. Reprint, Farnborough, England: Gregg International.

Vallet, Éric. 2010. *L'Arabie marchande: État et commerce sous les sultans rasūlides du Yémen (626–858/1229–1454)*. Paris: Publications de la Sorbonne.

Varisco, Daniel M. 1982. "The Adaptive Dynamics of Water Allocation in al-Ahjur, Yemen Arab Republic." PhD diss., University of Pennsylvania, Philadelphia (unpublished).

Varisco, Daniel M. 1983. "*Sayl* and *Ghayl*: The Ecology of Water Allocation in Yemen." *Human Ecology* 11:365–383.

Varisco, Daniel M. 1989. "Medieval Agricultural Texts from Rasulid Yemen." *Manuscripts of the Middle East* 4:150–154.

Varisco, Daniel M. 1991. "A Royal Crop Register from Rasulid Yemen." *Journal of the Economic and Social History of the Orient* 34:1–22.

Varisco, Daniel M. 1994. *Medieval Agriculture and Islamic Science: The Almanac of a Yemeni Sultan*. (Publications on the Near East, University of Washington, 6). Seattle: University of Washington Press.

Varisco, Daniel M. 2002. "Agriculture in Rasulid Zabīd." In *Studies on Arabia in Honour of Professor G. Rex Smith*, eds. John F. Healey and Venetia Porter, 323–351. (Journal of Semitic Studies, Supplement 14). Oxford: Oxford University Press.

Varisco, Daniel M. 2006. "The State of Agriculture in Late 13th Century Rasulid Yemen." In *Storia e cultura dello Yemen in età islamica: Con particolare riferimento al periodo Rasūlide. Convegno Roma, 30–31 ottobre 2003*, 161–174. (Accademia Nazionale dei Lincei, Fondazione Leone Caetani, 27). Rome: Bardi.

Varisco, Daniel M. 2009. "Agriculture in al-Hamdānī's Yemen: A Survey from Early Islamic Geographical Texts." *Journal of the Economic and Social History of the Orient* 52:382–412.

Varisco, Daniel M. 2010. "The *Milh al-Malâha* of al-Malik al-Ashraf ʿUmar (d. 696/1296): Situating the Ur-Text of the Rasulid Agricultural Corpus." *Chroniques du manuscrit au Yémen* 9. https://journals.openedition.org/cmy/1892.

Varisco, Daniel M. 2012. "*Qāt* and Traditional Healing in Yemen." *Herbal Medicine in Yemen: Traditional Knowledge and Practice, and Their Value for Today's World*, eds. Ingrid Hehmeyer and Hanne Schönig, with the collaboration of Anne Regourd, 69–102. (Islamic History and Civilization, 96). Leiden: Brill.

Varisco, Daniel M. 2015. "*Pars Pro Toto* Observation: Historical Anthropology in the Textual Field of Rasulid Yemen." *History and Anthropology* 26:92–109.

Varisco, Daniel M. and G. Rex Smith, eds. 1998. *The Manuscript of al-Malik al-Afḍal al-ʿAbbās b. ʿAlī b. Dāʾūd b. Yūsuf b. ʿUmar b. ʿAlī Ibn Rasūl (d. 778/1377): A Medieval Arabic Anthology from the Yemen*. Warminster: Aris and Phillips for the E. J. W. Gibb Memorial Trust.

Vera Chamaza, Galo W. 1994. "Der VIII. Feldzug Sargons II. Eine Untersuchung zu Politik und historischer Geographie des späten 8. Jhs. v. Chr. (Teil I)." *Archaeologische Mitteilungen aus Iran* 27:91–118.

Vera Chamaza, Galo W. 1995–1996. "Der VIII. Feldzug Sargons II. Eine Untersuchung zu Politik und historischer Geographie des späten 8. Jhs. v. Chr. (Teil II)." *Archaeologische Mitteilungen aus Iran* 28:235–267.

Vitruvius/Morgan, Morris H., trans. 1960. *The Ten Books on Architecture*. New York: Dover.

Vogt, Burkhard. 2004a. "Grundzüge der antiken südarabischen Bewässerungslandwirtschaft." *Beiträge zur allgemeinen und vergleichenden Archäologie* 24:67–104.

Vogt, Burkhard. 2004b. "Towards a New Dating of the Great Dam of Mārib. Preliminary Results of the 2002 Fieldwork of the German Institute of Archaeology." *Proceedings of the Seminar for Arabian Studies* 34:377–388.

Vogt, Burkhard. 2005. "The Great Dam, Eduardo Glaser and the Chronology of Ancient Irrigation in Ma'rib." In *Sabaean Studies: Archaeological, Epigraphical and Historical Studies in Honour of Yūsuf M. ʿAbdallāh, Alessandro de Maigret and Christian J. Robin on the Occasion of Their Sixtieth Birthdays*, eds. Amida M. Sholan, Sabina Antonini, and Mounir Arbach, 501–520. Naples and Ṣanʿāʾ: University of Ṣanʿāʾ, Yemeni-Italian Centre for Archaeological Researches, and Centre français d'archéologie et de sciences sociales de Ṣanʿāʾ.

Vogt, Burkhard. 2007. "Der aktuelle Forschungsstand zum Großen Damm von Marib, Jemen, nach der Winterkampagne 2005/2006." *Zeitschrift für Archäologie Außereuropäischer Kulturen* 2:121–142.

Warburg, Otto. 1916. *Die Pflanzenwelt*. Vol. 2, *Dicotyledonen, Vielfrüchtler (Polycarpiae) bis Kaktusartige Gewächse (Cactales)*. Leipzig and Vienna: Bibliographisches Institut.

Warren, D. Michael. 1996. "Comments on Article by Arun Agrawal." *Indigenous Knowledge and Development Monitor* 4 (1). http://app.iss.nl/ikdm/ikdm/ikdm/4-1/articles/agrawal.html.

Weber, Max. 2001. *Gesamtausgabe*. Abteilung 1, *Schriften und Reden*. Band 22, *Wirtschaft und Gesellschaft: Die Wirtschaft und die gesellschaftlichen Ordnungen und Mächte; Nachlass*. Teilband 2, *Religiöse Gemeinschaften*, ed. Hans G. Kippenberg, with the collaboration of Petra Schilm and Jutta Niemeier. Tübingen: Mohr Siebeck.

Wehr, Hans/Cowan, J. Milton, ed. 1979. *A Dictionary of Modern Written Arabic*. 4th ed. Wiesbaden: Harrassowitz.

Weiler, Elmar W., and Lutz Nover. 2008. *Allgemeine und molekulare Botanik (begründet von Wilhelm Nultsch)*. Stuttgart and New York: Thieme.

Wensinck, Arent J. (1927) 1960. *A Handbook of Early Muhammadan Tradition*. Reprint, Leiden: Brill.

Wiet, Gaston. (1932) 1984. *Catalogue général du Musée arabe du Caire: Objets en cuivre*. Reprint, Cairo: L'Organisation égyptienne générale du livre.

Wilkinson, John C. 1977. *Water and Tribal Settlement in South-East Arabia: A Study of the* Aflāj *of Oman*. (Oxford Research Studies in Geography). Oxford: Clarendon.

Wilkinson, Tony. 1999. "Settlement, Soil Erosion and Terraced Agriculture in Highland Yemen: A Preliminary Statement." *Proceedings of the Seminar for Arabian Studies* 29:183–191.

Wilkinson, Tony. 2002. "Agriculture and the Countryside." In *Queen of Sheba: Treasures from Ancient Yemen*, ed. St John Simpson, 102–107. London: The British Museum.

Wilkinson, Tony. 2006. "From Highland to Desert: The Organization of Landscape and Irrigation in Southern Arabia." In *Agricultural Strategies*, eds. Joyce Marcus and Charles Stanish, 38–68. Los Angeles: Cotsen Institute of Archaeology, University of California.

Winkler, Hans A. 1930. *Siegel und Charaktere in der muhammedanischen Zauberei*. (Studien zur Geschichte und Kultur des islamischen Orients, 7). Berlin and Leipzig: De Gruyter.

von Wissmann, Hermann/Müller, Walter W., ed. 1982. *Die Geschichte von Saba' II. Das Großreich der Sabäer bis zu seinem Ende im frühen 4. Jh. v. Chr.* (Österreichische Akademie der Wissenschaften, Philosophisch-historische Klasse, Sitzungsberichte, 402). Wien: Österreichische Akademie der Wissenschaften.

Witkam, Jan Just. 2007. "Gazing at the Sun: Remarks on the Egyptian Magician al-Būnī and His Work." In *O ye Gentlemen: Arabic Studies on Science and Literary Culture, in Honour of Remke Kruk*, eds. Arnoud Vrolijk and Jan P. Hogendijk, 183–199. Leiden: Brill.

Wöhrlin, Traugott. 1999. *Hölzerne Haustüren im Jemen*. Wiesbaden: Reichert.

Internet Sources

Social Fund for Development. n.d. "Background and Objectives." Accessed August 12, 2016. http://sfd.sfd-yemen.org/content/1/42.

Social Fund for Development. n.d. "Areas of Interventions [*sic*]." Accessed August 12, 2016. http://sfd.sfd-yemen.org/category/2.

Index

Terms that are included in the glossary are marked with an asterisk (*).
Page numbers for definitions or detailed explanations are in boldface.

www.ingramcontent.com/pod-product-compliance
Lightning Source LLC
LaVergne TN
LVHW010640300926
849161LV00005B/1004

* 9 7 8 9 0 0 4 3 8 7 0 1 0 *